# Practical Manual of HORTICULTURE

*The Author*

**Professor (Dr.) Kaushal Kumar Misra** received his Ph.D. Horticulture degree in the year 1986 from Govind Ballabh Pant University of Agriculture and Technology, Pantnagar-263145, U.K. He did his post doctorate degree from V. P. I. and State University, Blacksburg, Virginia, U.S.A. during 1989. He started his career as Senior Research Assistant on June 08, 1974 from Pantnagar. He was selected as Junior Research Officer, Horticulture on April 1, 1978. He was selected as Senior Research Officer / Associate Professor, Horticulture on April 1, 1991. Dr. Misra was promoted as Professor, Horticulture on July 27, 1998. He was Officer Incharge at Research Station, Sui, Lohaghat District Pithoragarh, U.P. from January 12, 1992 to March 31, 1994, then again as Associate Director/Joint Director at H.R.C., Patharchatta from November 19, 1997 to July 15, 1999. He was selected as Head of Department of Horticulture from November 1, 2004 to December 18, 2005 and September 30, 2006 to June 01, 2009. He published 21 books, 187 research papers, 150 popular articles and a dozon of book chapters in review books to his credit. He was awarded Dr. Rajendra Prasad Puruskar, 1984 from I.C.A.R., New Delhi, confered, ISHRD Fellowship in September, 2004 from Indian Society of Horticultural Research and Development, Uttarakhand, Best Citizen of India Awards, 2006 and 2010, New Delhi and fellowship from Horticultural Society of India award, 2013. Sixteen Ph.D. and fourteen M.Sc. (Ag.) Horticulture students received his/her degree under his guidance. He has got about 40 years of experience of teaching, research and extension education.

# Practical Manual of HORTICULTURE

— *Author* —

**Dr. Kaushal Kumar Misra**
*Ph.D., P.D. (U.S.A.),*
*Professor and Former Head*

*Department of Horticulture, College of Agriculture,*
*Govind Ballabh Pant University of Agriculture & Technology,*
*Pantnagar-263145, District – Udham Singh Nagar,*
*Uttarakhand, India*

**2024**
**BIOTECH BOOKS®**

ISBN 978-81-7622-354-6



*Published by*: **BIOTECH BOOKS®**
4762-63/23, Ansari Road, Darya Ganj,
New Delhi - 110 002
Phone: +91-011-23262132
E-mail: biotechbooks@yahoo.co.in

*Typeset at*: **Classic Computer Services**
Delhi - 110 035

*Printed at*: Replika Press Pvt. Ltd.

PRINTED IN INDIA

*Dedicated*

*to my Sister*

**Jagteswari Mishra (Awasthi)**

*and*

*to my brother*

**Shitansu Kumar Mishra**

# Preface

Undergraduate and graduate students of Horticulture do not have at present a "Practical Manual of Horticulture" of Indian Agricultural universities and colleges. The author is, therefore, making an attempt to meet this need. This is perhaps the only attempt so far made on the subject in India. It is hoped that manual will help to teachers and students alike. In all, the manual has fortiseven exercises covering diverse topics such as identification of garden tools, identification of horticultural crops, identification of vegetable crops, spices and condiments, identification of ornamental crops, medicinal and aromatic plants, preparation of seed bed for fruit plants, preparation of nursery bed for fruit plants, determination of thousand seed weight, testing of seed viability and germination, study of polyembryony in *Citrus jambheri*, presowing seed treatment for germination, practice of asexual methods of propagation: cutting and layering, practice of asexual methods of propagation: grafting, practice of asexual methods of propagation: budding, layout of an orchard, use of planting board for tree planting, planting of an orchard, training of fruit trees, pruning of fruit trees, yield estimate of an orchard, preparation of seed bed for vegetables and sowing of seeds, care of vegetable seedlings in the nursery and their transplanting, identification of ornamental plants, seedbed preparation and sowing of ornamental plants, potting and repotting of ornamental plants, training and pruning of ornamental plants, cultural practices in important ornamental plants, preparation of bonsai, planning and layout of gardens, preparation, planting and care of lawn, making of herbaceous border, making of shrubbery border, preparation of potting mixture, uses of flowers for different purposes, care and maintenance of green house/poly house plants, arranging flower show, designing of nursery experiments, media preparation, testing and potting of subtropical nursery plants and irrigation techniques, fertilization of subtropical nursery stock, testing of nursery soil and plant tissue of subtropical fruits, use of plant growth regulators in subtropical nursery production, protected nursery production of subtropical fruits, practice of *in-vitro* cloning of subtropical fruits – preparation of media and sterilization, selection and preparation of explants, establishment of explants and proliferation, stooling in guava and balling and burlapping of nursery plants. A list of common english names, botanical names and family of the various horticultural plants like fruits, vegetables, ornamental and landscaping, spices and condiments, plantation crops and medicinal and aromatic plants have been given in the appendices. The selected references presented at the end can be useful to obtain further information. In the last, glossary containing related terms have been given.

It is hoped that the teachers and students alike will welcome the manual and find it useful. The preparation of the manual involved considerable amount of hard work. In this task, I wish to express our sincere gratitude to Dr. Ranvir Singh, Ex-dean Agriculture and Professor of Horticulture, who took considerable amount of pain in this work. It is due to his preservance and consistent efforts that this manual has taken the present shape. Thanks are also due to Mrs. Manjul Mishra for her valuable help and constructive suggestions. I am thankful to my dauther Mrs. Kanan Kaushal Pandey, sons Mr. Praduman Kaushal, Mr. Panini Kaushal, my son-in-law Mr. Satyajeet Pandey and daughter-in-law Mrs. Pratibha Kaushal and my grand son Mr. Pratham Kumar. Thanks are also due to Mr. Shailendra Kumar, Senior Assistant for typing the manuscript on computer within the shortest possible time.

***Dr. Kaushal Kumar Misra***

# Contents


*Preface* — *vii*

1. Exercise No. 1 — 1
2. Exercise No. 2 — 3
3. Exercise No. 3 — 6
4. Exercise No. 4 — 9
5. Exercise No. 5 — 14
6. Exercise No. 6 — 16
7. Exercise No. 7 — 18
8. Exercise No. 8 — 20
9. Exercise No. 9 — 26
10. Exercise No. 10 — 28
11. Exercise No. 11 — 32
12. Exercise No. 12 — 36
13. Exercise No. 13 — 39
14. Exercise No. 14 — 41
15. Exercise No. 15 — 43
16. Exercise No. 16 — 46

17. Exercise No. 17 48

18. Exercise No. 18 50

19. Exercise No. 19 52

20. Exercise No. 20 55

21. Exercise No. 21 58

22. Exercise No. 22 60

23. Exercise No. 23 65

24. Exercise No. 24 68

25. Exercise No. 25 71

26. Exercise No. 26 74

27. Exercise No. 27 76

28. Exercise No. 28 79

29. Exercise No. 29 82

30. Exercise No. 30 87

31. Exercise No. 31 90

32. Exercise No. 32 92

33. Exercise No. 33 95

34. Exercise No. 34 98

35. Exercise No. 35 102

36. Exercise No. 36 105

37. Exercise No. 37 111

38. Exercise No. 38 115

39. Exercise No. 39 117

40. Exercise No. 40 119

41. Exercise No. 41 124

42. Exercise No. 42 129

43. Exercise No. 43 132

44. Exercise No. 44 134

45. Exercise No. 45 136

**46. Exercise No. 46** 138

**47. Exercise No. 47** 140

**Appendices**

**Appendix I** 145

**Appendix II** 148

**Appendix III** 151

**Appendix IV** 153

**Appendix V** 156

**Appendix VI** 157

**Appendix VII** 161

**Appendix VIII** 162

**Appendix IX** 165

**Appendix X** 167

**Glossary** 171

**References** 205

**Index** 209

# Exercise No. 1

## Objective

Identification of garden tools.

## Introduction

Garden tools are used for various purposes like nursery/seedbed preparation, sowing of seeds, weeding, spraying, grafting, budding, pruning, training etc., Identification of these tools is necessary for a student of Horticulture in order to know their use and handling and to avoid injury likely to occur in performing various operations. Identification and their proper use also increase the efficiency of operations and avoid wear and tear of the tools.

## Materials

Spade, Khurpi, Budding and Grafting knives, Pruning knife, Secateur, Pole pruner, rake, Lopper or Pruning shear, Hedge shear, Shovel, Can, Axe, Pruning saw, Hand Fork, Patch budder, Hand cultivator, Transplanting trowel, Plant Protector, Root irrigator, Dibbler, etc.

## Assignments

1. Draw the diagrams of various garden tools and lebel them properly.
2. Write the functions of each tool in the garden.

## Observation Blank

| *Sl.No.* | *Name of the Tool* | *Use* | *Precaution in Using of Tool* |
|---|---|---|---|

# Exercise No. 2

## Objective

Identification of horticultural crops.

## Introduction

Students of the horticulture are normally exposed to various types of horticultural crops like fruits, vegetables, ornamental and landscapes, medicinals and aromatics, spices and condiments and plantation crops. Identification of these crops is a must in order to acquaint the students about the behavior of the crops to which they will study and work. These crops include various types of plant species like annual, biennial, perennial, herbaceous woody, angiosperms, gymnosperms, ferns and higher group of fungi. The horticultural plants can be identified on the basis of their morphological characters like, shape, size, trunk, bark, branches, young shoots, spines, leaf petiole, blade, oilglands on the leaves and flower type, colour, pedicle length, calyx, corolla, androecium, gynoecium and fruit as the case may be. A sound knowledge of the structure and development of horticultural plants enable one to describe the plants accurately and to differentiate between the various plants.

## Materials and Equipments

Various types of fruits and plantation crops, scale, Vernier callipers, hand lens, tape, knife etc.

## Procedure

Various types of plant materials or their parts are either collected in the laboratory or studied *in situ*. The measurement of their plant parts are made and compared with each other. Also their colour, variations in plant parts like shape, number, presence or absence of certain specialized parts are noted visually or counted. The common english names, scientific or botanical names and family are also noted and the various plants may be arranged by grouping them as fruits, vegetables, ornamental and other kind of plants grouped in various categories in Horticulture. Then these plants are again arranged family wise in each horticultural group. The basis of identification may also be mentioned in each individual species which differentiate them from other species. The following information may be collected while identifying each species.

### A. General

1. Common name, botanical name, natural order, variety.
2. Rootstock, age, seedling or grafted.

### B. Vegetative Characters

Size of the plant, vigour, habit, crown shape, trunk, length and diameter, bark colour and surface, branching habit, young shoot characters like colour, hairiness, size, spine intensity and its size, colour and nature, leaf petiole length, breadth, presence and absence of wing, if present its shape, size and margin; leaf blade simple or compound, shape, nature of young and old leaves, presence and absence of oil glands with their density, shape and size.

### C. Flower Characters

Type of flowers, colour of flower buds, pedicle length; Calyx number, shape, colour, corolla number, shape, size, aestivation and roecium, arrangement, shape, size, number, position; gynoecium- stigma its shape, number and diameter, style its shape, length and diameter; ovary its type, diameter, shape, number of carpels, number of locules and colour; disc-shape, colour, diameter.

### D. Fruit Characters

Size, shape, peel colour, type, surface, apex, base, rind, adherence, presence or absence of oil glands on the skin, segments, vesicles size, shape, flesh colour, juicy or fleshy, seed-number, size, shape, cotyledon colour, chalaza colour, colour of inner seedcoat.

Record the observations in the following observation blank:

### Assignments

1. List various types of fruits giving their botanical names and family. Arrange fruits of a family at a place.
2. List various types of plantation crops with their botanical names and family.

| *Sl.No.* | *Common or English Name* | *Botanical Name* | *Family* |
|---|---|---|---|
| **A.** | **Fruits** | | |
| | a. Tropical | | |
| | b. Subtropical | | |
| | c. Temperate | | |
| **B.** | **Plantation crops** | | |

# Exercise No. 3

## Objective

Identification of vegetable crops, spices and condiments.

## Introduction

Vegetable crops, spices and condiments are the important components of horticultural crops. Their identification is a must for the students of Horticulture as they are to work with these crops. As in case of fruit and plantation crops, the identification of vegetables, spices and condiments can also be done by examining their plant parts like roots, stem, leaves, flowers, fruits and seeds.

## Materials

Various types of vegetables, spices and condiments, scale, Vernier callipers, Hand lens, knife, pencil etc.

## Procedure

Same as in case of fruit and plantation crops.

## Assignments

1. List various types of vegetables, spices and condiments with their botanical names and family.

## Observation Blank

| *Sl.No.* | *Common Name* | *Botanical Name* | *Family* |
|---|---|---|---|
| **A.** | **Fruits** | | |
| 1. | Vegetables | | |
| | a. Cole crops | | |
| | b. Bulb crops | | |
| | c. Cucurbits | | |
| | d. Fruits vegetables | | |
| | e. Leafy vegetables | | |
| | f. Leguminous vegetables | | |
| | g. Root vegetables | | |

| Sl.No. | Common Name | Botanical Name | Family |
|---|---|---|---|
| | h. Tuber crops | | |
| | i. Salad crops | | |
| | j. Miscellaneous crops | | |
| 2. | Spices and condiments | | |
| | a. Spices | | |
| | b. Condiments | | |

# Exercise No. 4

## Objective

Identification of ornamental crops, medicinal and aromatic plants.

## Introduction

These are a large number of ornamental plants belonging to several species under various families. These plants have been classified as annuals, biennials, shrubs, trees, cacti and succulents, ferns, climbers, palms and cycades, orchids, lawn and turf grasses, water and marshyland plants, mangroves and rock plants. Similarly, there are large number of medicinal and aromatic plants belonging to several diverse species and families. Students of Horticulture are to work with these plants. Therefore, a systematic knowledge of their growth behavior, flowering and fruiting is essential.

## Materials

Various types of ornamental, medicinal and aromatic plants, scale, Vernier callipers, Hand lens, Knife, Pencil etc.

## Procedure

Same as in case of fruit and plantation crops.

## Assignments

1. List various types of ornamental and landscape plants with their botanical names and family.
2. List various types of medicinal and aromatic plants with their botanical names and family.

**Observation Blank**

| Sl.No. | Common Name | Botanical Name | Family |
|---|---|---|---|
| 1. | Ornamental crops | | |
| | a. Annuals | | |
| | i) Summer | | |
| | ii) Winter | | |
| | b. Biennial | | |
| | c. Shrubs | | |
| | i) Flowering shrubs | | |
| | ii) Foliage shrubs | | |
| | iii) Hedge plants | | |
| | iv) Edge plants | | |

| Sl.No. | Common Name | Botanical Name | Family |
|---|---|---|---|
| | v) Climbers | | |
| | d. Trees | | |
| | i) Flowering | | |
| | ii) Avenue | | |
| | e. Cactii and Succulents | | |
| | i) Cactii | | |
| | ii) Succulents | | |
| | f. Ferns | | |
| | g. Palms and cycades | | |
| | i) Palms | | |

| *Sl.No.* | *Common Name* | *Botanical Name* | *Family* |
|---|---|---|---|
| | ii) Cycades | | |
| | h. Orchids | | |
| | i. Lawn and Turf grasses | | |
| | i) Lawn grasses | | |
| | ii) Turf grasses | | |
| | j. Water and Marshyland plants | | |
| | i) Water plants | | |
| | ii) Marshyland plants | | |
| | k. Mangroves | | |

| Sl.No. | Common Name | Botanical Name | Family |
|---|---|---|---|
| | 1. Rock plants | | |
| 2. | Medicinal and aromatic plants | | |
| | i) Medicinal plants | | |
| | ii) Aromatic plants | | |

# Exercise No. 5

## Objective

Preparation of seed bed for fruit plants.

## Introduction

Nurserymen generally use to prepare seedbeds for sowing of seeds. A proper seed bed may determine the germination of seeds and further seedling growth in seed bed. A good seed bed should have a loose but fine physical texture that produces close contact between seed and soil so that the moisture can be supplied continuously to the seed besides providing good aeration but not dries too rapidly. The surface soil should have the texture that may not form a crust. The subsoil should be permeable to air and moisture with good drainage and aeration. The soil should hold sufficient moisture but should not be waterlogged or anaerobic. A medium loam textured soil is considered better in this regard. A good soil of seedbed is one which has three fourth of the soil particles in the range of 1-12 mm in diameter.

## Materials

Rope, Spade, Khurpi, Measuring tape, Compost or F.Y.M., Rake.

## Procedure

The method of seedbed preparation depends upon the size of operation and the crop being grown. The first operation for this is to spade the soil to a depth of 15-20 cm and then clods are broken with the help of the spade if the area is small otherwise planking may be done in case of larger area. The soil is raked vigorously and weeds are removed. The soil is leveled and manures are added and mixed thoroughly in the upper layer of 10 cm. Drainage and irrigation channels are prepared. The seedbeds are prepared by raising the soil to about 15 cm height. The width of the bed should be kept about 1.25 metres so that weeding operation can be performed easily. There is no limit of length of the bed. About 60 cm space is left between the two beds which is used for walking and as an irrigation channel. Before sowing of seeds in the bed, mixing of insecticide may be needed to kill the insects in the bed. If the soil is infected with some pathogens and nematodes then the soil must be fumigated before using it for nursery.

**Assignments**

1. Prepare the seedbed for sowing of citrus seeds.

# Exercise No. 6

## Objective

Preparation of nurserybed for fruit plants.

## Introduction

Nurserybeds are the bigger sized beds in which the seedlings of the seedbeds are planted at a wider distance where they are kept for one or two years till they are planted in the orchard or sold. The distance between rows and plants depends upon fruit species and the duration to which seedlings are to be grown. A good nursery bed should have a loose and fine physical texture in order to provide a good aeration and hold sufficient moisture. The soil should be well drained. A clay loam textured soil is considered where the plants are required to be uprooted with earth ball while light textured soils like sandy loam to loam are required where the plants are required to be removed bareroot. Nursery beds should be located in an open area and near the water source.

## Materials

Rope, Spade, Khurpi, Measuring tape, Compost or F.Y.M., Rake.

## Procedure

Nursery beds should be laidout in such a way that there is an access to all the beds through roads and paths. It is better to divide nursery beds into sections based on fruit crops say one section for mango, other for citrus fruits and so on. Mainly nursery beds for fruit plants are prepared on flate lands after giving proper slope to drain the excess water. Square or rectangular beds may be prepared depending upon the choice of the nurserymen, ease of operation, species to be grown, irrigation and other facilities like availability of land, orientation etc.

The method of nurserybed preparation will depend upon the size of operation. The first operation for this is to spade or plough the soil to a depth of 15-20 cm and then clods are broken with the help of the spade if the area is small otherwise planking may be done in case of larger area. The soil is raked vigorously and weeds are removed. The plot is leveled. Drainage and irrigation channels are prepared. Paths and roads are made. The beds are prepared and leveled and the manures are added in the soil of the beds thoroughly in the upper layer of 10-15 cm. If the soil is infected, insecticide may be added to the soil before planting. If the soil is infected with nematodes or any pathogen then nurserybeds must be fumigated before their use.

**Assignments**

1. Prepare the nurserybed for transplanting of six month old rootstock seedlings of citrus or for one year old sproutd and rooted cuttings of lemon.

# Exercise No. 7

## Objective

Determination of thousand seed weight.

## Introduction

Test is conducted to know the boldness and soundness of seeds. The weight of a given number of seeds of different varieties varies. This variation is due to boldness of the seeds. The boldness and soundness of the seeds shall be vary due to the soil fertility, varieties and environmental conditions. Higher the seed weight of a variety per thousand seeds, better is the variety for cultivation.

## Materials and Equipment

Scale, balance and seeds.

## Procedure

Draw the sample of the seeds uniformly, mix them thoroughly, count 3 to 4 lots each of hundred seeds for each variety. Note the weight of each lot separately as shown in the observation blank.

**Observation Blank**

Crop - Variety-

| *Lots* | *Variety* | | | | *Weight (g)* |
|---|---|---|---|---|---|
| | *1* | 2 | 3 | 4 | *Mean* |
| 1 | | | | | |
| 2 | | | | | |
| 3 | | | | | |
| 4 | | | | | |
| Total | | | | | |

**Conclusion**

Seed sample having bold seeds always record higher weight per thousand seeds.

**Assessment**

1. Find out the thousand seed weight of different varieties of okra, radish, pea and cauliflower.
2. Find the thousand seeds weight of lime, lemon, grapefruit, mandarin and pummelo.

# Exercise No. 8

**Objective**

Testing of seed viability and germination.

**Introduction**

The viability test shows the viability of the embryo. A viable embryo may or may not give rise to a healthy seedlings. Germination test gives the development of healthy seedlings. Germination and viability tests prove useful in adjusting the seed rate. Sowing seeds in the nursery bed or in the field without conducting test results in loss of seeds. Germination tests are very important for vegetable crops, flowering trees, annuals and for several fruits.

The following precautions are needed in order to ensure good results. 1. Use large seed sample and unbiased sample (100-200 seeds), 2. Provide optimum and uniform spacing between seeds, 3. Ensure proper aeration and moisture in the germinating, 4. Sterlize all the glassware and equipments before use to avoid growth of micro-organisms.

1. Germination tests:
   (i) Covered petri dish test.
   (ii) Rolled towel test.
2. Varibility tests:
   (i) Excised embryo test.
   (ii) T T C test.
   (iii) Indocarmine test.

## 1. Germination Test

### (i) Covered Petri Dish Test

### Materials and Equipments

Petri dishes, blotting papers (white), tally counter, labels and scissor.

### Procedure

Take 4 pairs of petri dishes, cut blotting paper of the same size of petri dishes. Place 3-4 layers of blotting paper in each dish. Mix the seeds thoroughly count 100 seeds to be tested. Divide the seeds randomly in 4 lots each of 100 seeds. Place 100 seeds in each petri dish at uniform distance. Moisten the blotting paper and cover the petri dish. Keep the petri dishes in warm place. Keep the blotting paper moist by sprinkling water as and when necessary. Do not allow the blotting papers to dry. Count and remove the sprouted seeds at every alternate day. If any two of the lots differ by more than ten per cent a retest should be carried out. Note the data as outlined below:

### Observation Recorded

Variety - Temperature - Date-

| *Lot* | *No. of Seeds at Start* | *No. of Seeds Germinated (Days)* | | | | | | *Total No. of Seeds Germinated* | *Percentage germination* |
|---|---|---|---|---|---|---|---|---|---|
| | | *2nd* | *4th* | *6th* | *8th* | *10th* | *12th* | | |
| 1 | | | | | | | | | |
| 2 | | | | | | | | | |
| 3 | | | | | | | | | |
| 4 | | | | | | | | | |

### (ii) Rolled Towel Test

### Materials and Equipment

Blotting paper (20×30 cm.).

### Procedure

Take a large sized (20×30 cm.) blotting paper. Moisten the blotting paper. Place the seeds well spaced along on side of the paper in such a manner that the edge of the blotting paper can cover the seed. Rolled the blotting paper and hold it in this position. Place another row of seeds and roll the blotting paper again. This way keep the seeds in 4 to 6 layers. Each time rolling should be light. Place the rolled blotting paper in a suitable place for germination. Keep the blotting paper moist. Record the data as given above.

## 2. Viability Tests

Viability is represented by germination percentage which expresses the number of seedlings which can be produced by a given number of seeds. Germination should be prompt and growth of seedlings should be vigorous. This is the seed viability or germinating power and may be represented by germination rate. In seeds of low viability, a low germination percentage and low germination rate are often associated. A reduction in the seed viability and seed vitality may result from incomplete seed development on the plant, injuries during harvest, improper processing and storage or ageing.

### Measurement of Viability

If one measures the time sequence of germination of a given lot of seeds or the emergence of seedlings from a seedbed one usually finds a pattern that is illustrated by the germination curb.

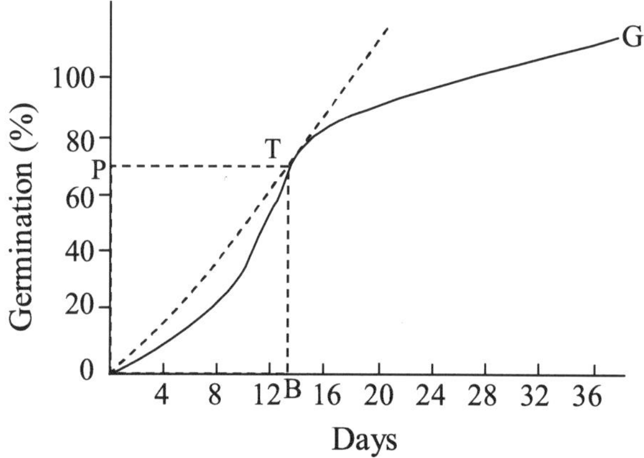


**Figure 1: Typical Germination Curve of a Seed Sample**

There is an initial delay in the start of germination, then a rapid increase in the number of seeds that germinate followed by a decrease in the rate of appearance. When viability is less than 100 per cent, the exact end point may be difficult to ascertain.

Measurement of germination involves two factors, the germination percentage and the germination rate. Germination percentage should involve a time element indicating the number of seedlings produced within a specified length of time. Germination rate can be measured by several methods. One determines the number of days required to produce a given germination percentage. A better method that has been used for many years. Calculates the average number of days required for radicle or plumule emergence (Elemion 1938; Harrington, 1963).

$$\text{Mean days} = \frac{N_1 T_1 + N_2 T_2 + \text{----} + N_X T_X}{\text{Total number of seeds germinating}}$$

N values are the number of seeds germinating within consecutive intervals of time. T values indicate the time between the beginning of the test and end of the particular interval of measurement. Katowski (1926) has used the reciprocal of this formula multiplied by 100 to determine a coefficient of velocity Czabator (1962) has suggested another measurement that includes both rate and percentage germination value (GV). The calculate GV, a germination curb must be obtained by periodic counts of radicle or plumule emergence. The important values on the curve are T-the point at which the germination rate begins to slow down and G-The final germination percentage.

These points divide the curve into two parts – a rapid phase and a slow phase. P V (Peak value) is the germination percentage at T divided by the days to reach that point. Mean daily germination (M D G) is the final germination percentage devided by number of days in the test.

Then G V (Germination value) = P V (Peak value) × M D G (Mean daily germination)

**(i) Excised Embryo Test**

**Materials and Equipments**

As for covered petri dish test.

**Procedure**

This method is very effective for seeds having prolonged dormancy. Take 25 seeds and remove the embryo from the cotyledon or endosperm. Placed the naked embryo on sterilized nutrient agar media and watch the germination. Note the observation.

**(ii) Triphenyl Tetrazolium Chloride Test (T T C test)**

**Introduction**

The tetrazolium test is a biochemical method in which viability is tested not by germination but by red colour appearing when the seeds are soaked in 2, 3, 5 – triphenyl tetrazolium chloride. This chemical is absorbed with cells of living tissue where it is changed to an insoluble red compound formazan. Non living tissue remains uncoloured. The test was developed by Lakon (1949) who referred to it as a topographical test since loss in embryo viability begins to appear at the extremity of the radicle, epicotyls and cotyledon tips. The reaction takes place equally well in dormant and nondormant seeds. Results can be obtained with 24 hours and often less time. The T T C method has been used as a rapid approximate test of viability or as a germination test of dormant seeds that do not respond to other methods. The chemical is soluble in water making a colourless solution. Although the solution deteriorates with exposure to light, it will remain in good condition by several months if protected. It should be discarded if it becomes yellowish, an 1 per cent solution is commonly used, although somewhat lower concentrations of 0.05 per cent have been satisfactory.

## Materials and Equipments

Petri dishes, blotting paper and T T C.

## Procedure

Any hard covering such as an endocarp wing or scale must be removed. Seeds should be soaked in the dark. They must be moistened to acivate emzymes and to facilitate removal of seed coverings. Most seeds require preparation for T T C absorption. Some embryos with cotyledons such as prunes, apple and pear are excised completely. Seeds are soaked in T T C solution for 2-24 hours. Cut the seeds require a shorter time, those with exposed embryos somewhat longer intact seeds 24 hours or more. Interpretation of results depends upon the kind of seeds. Completely coloured embryos indicate good seeds. Conifers and other seeds with substantial endosperm must have both endosperm and embryo stained. In grass and grain seeds only the embryo colours.

## Observation to Taken

Crop name- Date-

| *Lot No.* | *No. of Seeds for T T C Treatement* | *No. of Seeds for Water Treatment* | *No. of Seeds Stained with T T C* | *No. of Seeds Germinated in Petri Dish* | *Per cent Seeds Stained* | *Per cent Germinated* |
|---|---|---|---|---|---|---|
| 1 | | | | | | |
| 2 | | | | | | |
| 3 | | | | | | |
| 4 | | | | | | |

## (iii) Indocarmine Test

## Materials and Equipements

Indocarmine dye and test tubes.

## Procedure

In this test the dead seeds get stained with dye due to increased permeability. Known number of seeds be taken in a test tube (25 seeds). Add sufficient dye solution to cover the seeds. The seeds which are dead get stained while the living ones remain unchanged. Better results are obtained when seed coat of seeds is removed.

$$\text{Real value of seeds} = \frac{\text{Purity percentage} \times \text{germination percentage}}{100}$$

## Assignments

1. Study the germination percentage of papaya, radish, carrot and tomato seeds by different methods.
2. Treat the seeds of different species of citrus with tetrazolium chloride and compare the result with regular germination.
3. Study the germination of fresh and one year old seeds of different fruits and vegetables purchased from seed stores.

# Exercise No. 9

## Objective

Study of polyembryony in *Citrus jambhiri*.

## Introduction

When two or more embryos are found within a seed, the phenomenon is known as polyembryony. In many species of genus Citrus, Sizigiam, Mangifera seeds contain two kind of embryos, sexual and asexual. The sexual embryo develops of the fusion of male and female gametes. The asexual embryo develops from the vegetative parts of the mother cell. The phenomenon by which the asexual embryos develop is known as apomixis. The seedlings develop from the asexual embryos are known as apomicts and seedlings as apogemetic seedlings. These seedlings are uniform and as good as plants grown by other vegetative means. The asexual embryo may develop by following methods.

1. Recurrent apomixis, 2. Adventitious embryony, 3. Non-recurrent and 4. Vegetative apomixis.

The number of asexual embryo within the seed varies from 2-20. The asexual embryos are usually larger in size and vigorous than sexual embryo.

## Materials and Equipments

Watch glass, needle, forecep, beaker, hand lens, Verniar callipers, dissecting microscope and jambhiri seeds.

## Procedure

Take 100 freshly extracted jambhiri seeds. Soak them in a beaker containing plain water for 4 to 6 hours. Hold the seeds by finger and remove the seed coat and testa with the help of a forcep. Open the seeds with the help of needle in two separate the cotyledons. Look for the embryos, which attached with cotyledons. The asexual and sexual embryos with the help of needle. Use a hand lens for separating the large and small embryo. Measure the size and large embryo with the help of Vernier calliper. Record the data as given below. Dissecting microscope be used for measuring the size of embryos.

## Observations Recorded

Variety name - ______________ Date________________

| *Number of Seeds Examined* | *No. of Embryos* | | *Size of Embryos* | | *Average Size (mm)* | | |
|---|---|---|---|---|---|---|---|
| | *Large* | *Small* | *Large (mm)* | *Small (mm)* | *Large* | *Small* | *Polyembryony (per cent)* |
| 1 | | | | | | | |
| 2 | | | | | | | |
| 3 | | | | | | | |
| 4 | | | | | | | |
| Total | | | | | | | |
| Mean | | | | | | | |

## Assignments

1. Study the extent of polyembryony in acid lime, grapefruit, mandarin, jamun and mango.

# Exercise No. 10

## Objective

Presowing seed treatment for germination.

## Introduction

Most of the types of seeds will germination getting suitable conditions for their germination. There are some species which need some special treatments to overcome their shortcomings like after ripening, dormancy and hard coatedness. Nature have provided some special mechanism for these seeds to prolong their life span since they are to pass through adverse climatic conditions. Under natural conditions, seeds may have to face some unformable conditions but to get suitable germination. They are given some special treatments to protect them against fungus, insects, that attack during storage. The aims of treating the seeds to provide them long life and protection against unfavourable conditions. Seed treatments are usually given prior to sowing and hence are known as pretreatments.

## Materials

Seeds, water, acid, temperature below 7°C.

## Procedure

### Cold Water Treatment

Seeds of some plant species like guava, arjun, jamun, neem, papaya and custard apple need to put in the clean water for 24 hours for rapid germination. Floating seeds on the water surface are usually immature or damaged by insects, diseases, or by thrashing. Such seeds are discarded. The water is drained off and the seeds are dried under shade to make them to retain minimum moisture. Seeds having hard coats do not germinate easily. To make the water enter into the seeds and reach the embryo, the seeds are immersed in 24 hours. Seeds of some vegetables and forest species are treated like this.

### Hot Water Treatment

Water at 40 to 50°C is used to expedit the germination and to arrest the growth of pathogens sticking to seed coat. Seeds of acacia, gulmohar and subabul etc. are immersed in boiling water for an hour or upto 48 hours. The seeds coming to the top of water are floated off and are discarded. Hard coated seeds are given this treatment.

### Alternate Drying and Wetting Treatment

In some cases where the hard seed coat is tough and can not be broken or cracked either with hot or cold water, seeds are soaked in water for 3-4 days and then dried and against soaked in water for 3-4 days and this cycle is repeated till the seeds show some signs of their germination. Seeds which swell have absorbed water and are able to germinate. In case of teak, these seeds are kept in such condition for about 45 days.

### Acid Treatment

Hard coated seeds of guava are made to germinate after immersing them in 20 per cent sulphuric acid for 20 minutes or in concentrated sulphuric acid for 1 to 5 minutes. Put acid in water. Acid should be added slowly and then the solution may be mixed with a glass rod. For this treatment, earthen pot and glass pot may be used. Seeds should not be stirred in the acid with hand. Wash the seeds with clean water atleast twice. No trace of acid should be sticking to the seed. Seeds after washing may be dried under shade by spreading them over a mat to drain off the excessive water. Treated seeds should be sown soon. Nitric acid, hydrochloric and other acids have been tried for breaking the dormancy with success to overcome the hard-coatedness of seeds.

### Other Chemical Treatments

Alkalies such as sodium hydroxide or Potassium hydroxide have corrosive action on the seed coat which makes it easier to imbibe moisture. These chemicals remove these structures like hair and fuzz and help in easy placement of the seed for sowing. Seeds treated with chemicals are sown soon after being treated and should not be stored.

### Scarification

Seeds with hard seed coats that are impermeable to water are scarified either by physical, chemical or mechanical means. The time of shaking may vary with the spead of shaking. Generally the seeds shaken for 20 minutes. Scarification has to be done with little or no damage to the seeds. Seeds need this type of scarification are crotolaria, asparagus, okra, Ber and guava. Mechanical scarifiers allow the seeds to pass over abrasive surface which may be stone or a sand paper or a disc rotated in such way that the seeds are rubbed fully.

## Stratification

Seeds of temperate fruits like apple, pear, peach, cherry, almond and walnut and some forest species need some period of rest under chilling conditions. Temperature below 7°C for about 60 days or so depending upon the species. Stratification may be done in two ways.

### (i) Indoor Stratification

Seeds immersed in moist sand, kept loose and then sealed in a polythene bag are placed in a fridge. Small quantity of seed can be stratified in this way. Sufficient quantity of sand with ample moisture has to be kept in the bag. The temperature of the chamber should be below 7°C throughout the period of seed keeping. After taking the polythene bag out of the refrigerator, it may be kept outside to bring the contents to room temperature. Later, the seeds are sown in the field or in the nursery.

### (ii) Field Stratification

Seeds of the desired species are placed in the field. A trench of the width of 30cm and depth of 30cm is dug at a convenient place in the field. A layer of sand, 5-6cm deept is placed at the bottom of the trench and then seeds are spread uniformly in a layer. The layer of seeds is covered with a layer of sand which may be 6cm or so thick. Another seed layer can again be put on the second layer of sand. In this way, the seeds are in between the two layers of sand which are 5-6cm thick. Ample moisture is provided to the sand layers so that the seeds do not get desiccated during stratification. Top layer is covered with a soil layer. The seeds remain in the trench for a specified period. The period of stratification will depend upon the kind of species and prevailing conditions. After stratification, the seeds are taken out and sown in the nursery.

## Observation Blank

Date:.........................

| *Sl.No.* | *Name of Variety* | *Number of Seeds* | *Treatment* | *Number of Seeds Germinated (Days after)* | | | | | | *Total* | *Per cent Germination* | *Remark* |
|---|---|---|---|---|---|---|---|---|---|---|---|---|
| | | | | *2* | *4* | *6* | *8* | *10* | *12* | | | |
| 1. | | | | | | | | | | | | |
| 2. | | | | | | | | | | | | |
| 3. | | | | | | | | | | | | |
| 4. | | | | | | | | | | | | |
| 5. | | | | | | | | | | | | |
| Total | | | | | | | | | | | | |

**Assignments**

1. Calculate the hours required for hot water treatment of seeds of acacia, gulmohar and subabul?
2. Calculate the days required for seeds of low chill apple for stratification?
3. Calculate the minutes required for scarification of seeds of Ber and guava.

# Exercise No. 11

## Objective

Practice of asexual methods of propagation: cutting and layering.

## Introduction

In order to obtain true to the type plants, asexual method of propagation is employed for various horticultural plants. Different method of asexual propagation like use of apomictic seedlings, cutting, layering, budding and grafting are used for propagation of various species and varieties of fruit, ornamental and plantation crops. Apomictic seedlings are used for multiplication of true to type rootstocks in case of citrus and mango while other methods are used for propagation of variety of plants. Cuttings may be defined as rooting of a piece of stem and leaf bud and regeneration of both a new shoot system from adventitious bud as well as new adventitious roots in the pieces of roots and leaves. The success in the multiplication of stem cuttings depends upon factors such as condition of the mother plants, position of the tree, time of the year, precaution in planting and after care. Cuttings can be made during rainy season (July-Sept) and spring (Feb-March) depending upon the plant species. Layering includes several forms of ground and aerial layering. When rooting is encouraged on the aerial parts of the plant after wounding it is known as air layering and when branches running parallel to ground are utilized, it is known as ground layering. Layering is the development of the roots on the stem while it is still attached to the mother plant.

## Materials

Mother plant, Secateur, Grafting knife, Moss grass, Plastic strips.

## Procedure

### A. Propagation by Cutting

Cuttings can be classified as stem cutting, root cutting and leaf cutting. Stem cuttings can be classified as hardwood, semi hardwood, soft wood and herbaceous cuttings. Hardwood cuttings are those made of matured, dormant, firm wood after leaves have abscised. The branches of current year or past season growth having about lead pencil thickness are collected from healthy, vigorous and young plants, Cuttings of 20-25 cm length are prepared. The basal cut is given about 0.3 cm below a bud. The leaves are removed from the cuttings. Holes are made in the rooting media at 2.5 cm apart and 5 cm between rows. Cuttings are planted in these holes by inserting two third basal part of the cutting in the holes. Soil around the cuttings is firmly pressed and watering is done if the rooting media is dry.

Semi hardwood cuttings are made from semi mature branches from the selected trees. The size of the cuttings is kept in between 10-15 cm. the leaves from the 2-3 basal part of cuttings are removed and the cuts are given as in case of hardwood cuttings. Such cuttings are required to be planted under mist. Softwood cuttings are prepared out of immature branches from the selected plants. These branches should be of 4-6 months old. The leaves are removed only from the basal part of the cutting which is to be buried inside the rooting media. Only the branches growing in full sunlight should be selected for preparing such cuttings. The basal cut is given just below the node and the cuttings are planted in mist chamber for rooting. The herbaceous cuttings are also prepared and about 10 cm. Herbaceous cuttings are made from succulent non-woody plants. Leaf cuttings are prepared from the leaf blade and petiole. The long leaves are cut into the sections of 8-10 cm length and these leaf pieces are inserted three-fourths of their length into the rooting medium. In flashy leaves, the large veins are cut on the under surface of the mature leaves which is then laid flate on the surface of the propagating medium. Root cuttings are taken from young stock plants in late winter or early spring but before new growth starts. It is essential to maintain the correct polarity while planting root cuttings. To avoid planting them upside down, the proximal end may be made with a straight cut and the distal end a slanting cut. Certain plants are difficult to root and can not root without any hormonal treatments.

### Observation Blank

| *Variety* | *Date of Planting* | *Number of Cuttings Planted* | *Time Required for Sprouting (days)* | | | | *Number of Cuttings Sprouted* | *Average Number of Roots per Cutting* |
|---|---|---|---|---|---|---|---|---|
| | | | *10* | *15* | *20* | *25* | | |
| a. Hardwood | | | | | | | | |

| *Variety* | *Date of Planting* | *Number of Cuttings Planted* | *Time Required for Sprouting (days)* | | | | *Number of Cuttings Sprouted* | *Average Number of Roots per Cutting* |
|---|---|---|---|---|---|---|---|---|
| | | | *10* | *15* | *20* | *25* | | |
| b. Semi hardwood | | | | | | | | |
| c. Softwood | | | | | | | | |
| d. Herbaceous | | | | | | | | |
| 2. Root Cutting | | | | | | | | |
| 3. Leaf cutting | | | | | | | | |

## B. Layering

### 1. Air-layering

One or two year old branch of about a lead pencil thickness is selected. This branch is wounded by girdling just below a node about 25-35 cm away from the growing tips. The bark of the girdled area (about 2.5-3.5 cm) is removed. A handful of moist moss grass is put around the girdled area and carefully wrapped with a piece of polythene over the moss grass completely. Both the ends are tied tightly when roots develop in the layers which can be easily seen through the plastic film, separation of the air layers is done. This stage reaches after 45-60 days of operation. After detaching the rooted layers, these are planted in semi shady place. Before planting some of the foliage is also removed for better establishment.

## 2. Simple Ground Layering

The cane tip of the plant is bend down to the ground to be multiplied. A ring of bark is removed at 20-25 cm away from the growing tip. The ground soil is removed up to a depth of 10 cm and injured cane portion is buried in the soil after placing some more soil. The soil is kept moist. The roots will develop after 4-6 weeks. The rooted canes are removed and planted in cool shady places.

### Observation Blank

| *Variety* | *Date of Layering* | *Number of Layers made* | *Date of Separation* | *Number of Layers Rooted* | *Av. Number of Roots per Layers* | *Av. Length of Roots* |
|---|---|---|---|---|---|---|
| | | | | | | |

### Assignments

1. Prepare various types of cuttings
2. Practice the air layering in litchi

# Exercise No. 12

## Objective

Practice of asexual methods of propagation: grafting.

## Introduction

The term "grafting" includes all forms of grafting and budding. It is an art of joining two plant parts together in such a manner that they will unite and continue their growth as one plant. A grafted plant consists of two entities, a rootstock and a scion. The part of the graft combination which give rise the top of the plant is known as scion and the part which supports the top is known as rootstock. When the scion part is a branch containing more than one bud, the operation is known as "grafting". Like budding, grafting is also successful within the close relatives. Better success is obtained when the rootstock and scion are equal in thickness and the operation is performed during rainy, spring or winter season depending upon the species and method of grafting. In evergreen species, rainy and spring seasons are better but in deciduous species, grafting is performed during winter season. The grafting makes possible to multiply the varieties that can not be multiplied by other methods of asexual propagation and the change of top to get the benefits of interstock.

## Materials

Rootstock, Scion, Mother plant, Secateur, Grafting and budding knife, Plastic strips.

## Procedure

There are two types of grafting *i.e.*, attached and detached grafting. In attached grafting, the main methods are inarching and tongue approach grafting. In detached grafting, the important methods are whip, tongue, cleft, side and veneer grafting. In spliced approached grafting, the stock plant of 1.5 – 2 years of age is selected and placed along the side of the mother plant. The branches of the same thickness are selected in such a way so that the scion branch comes parallel to the stock plant. The stock and scion branches are brought together and a thin layer of bark along with the wood is removed from 3.5 cm area of the stock and scion and both cuts are joined together in such a way that cambium layers meet easily. The cut surfaces are tied together without leaving any space between them. The tying is normally done with plastic film strips. After 6-8 weeks when union has completed, the stock is cut above the grafted portion and scion below it. The tongue approach grafting differs from the simple approach grafting in the manner that a tongue is made on both stock and scion, for better cambial contact.

In veneer grafting, 10-12 months old stock of 1-2 cm diameter is selected on which a shallow longitudinal cut of about 3 cm is made on one side of the stock plant at a height of about 10-15 cm from the soil surface with the help of a sharp knife. The bark is removed alongwith wood making an oblique cut. The cut should not be deeper than one third of the thickness of the stock at the lower end of the slanting cut. The branches of the desired varieties with 10-15 cm length and a lead pencil thickness are selected and the leaves of the selected branches are removed while still attached to the mother plant. It is done about 8-10 days before the actual date of grafting. After 8-10 days, the defoliated branches are separated from the mother plants for grafting operation. On the basal end of the scion, an identical slanting cut is made as that on the stock plant. A small cut is also given on the other side of the scion. The scion is fitted with the cut part of the stock matching cambium closely and the grafted portion is tied with the help of a 1.25 cm wide plastic strip leaving the terminal end free. Side grafting is similar to the veneer grafting except that only inner cut is given on the scion and a single cut on the stock. In case of cleft grafting the stock plant is headed back at a height of 20-25 cm from the soil surface with the help of a secateurs. In the centre of the stock a "V" shape cut is made with the help of a sharp knife. The scion is defoliated as in case of veneer grafting and collected on the day of operation. A wedge shaped cut is made at the lower end of the scion of the similar size as in case of stock. The wedge of the scion is inserted in the "V" shaped cut of the stock carefully in order to bring the cambium of both stock and scion in close contact. The grafted portion is then tied with the plastic strip leaving the terminal end free. When the union is completed and scion starts growing, the plastic strip is removed in order to avoid girdling.

**Observation Blank**

| *Sl.No.* | *Fruit Plant* | *Suitable Grafting Method* |
|---|---|---|
| 1. | Mango | |
| 2. | Guava | |
| 3. | | |
| 4. | | |
| 5. | | |
| 6. | | |

**Assignments**

1. Practice various types of grafting and review the various methods of the grafting suitable for various fruit plants.

# Exercise No. 13

## Objective

Practice of asexual methods of propagation: budding.

## Introduction

A budded plant consists of two entities, a rootstock or understock or stock and a scion or cion. The part of the graft combination which will give rise the top of the plant is known as scion and the part which support the top *i.e.,* roots is known as rootstock. When the scion is a piece of bark containing only one bud, the operation is termed as "budding". The budding is successful within the close relatives. Better results are obtained when the rootstock is one or two year old and in active growth. Spring season is better for evergreen plants. There are several common methods of budding *i.e.,* shield or T-budding, Patch budding and Chip budding which are used in multiplication of plants. The choice of method depends on the species. Budding can also be performed when scion material is in shortage.

## Materials

Rootstock, Scion, Mother plant, Secateur, Grafting and budding knife, Plastic strips.

## Procedure

In shield budding a healthy uniform stock plants of about a lead pencil thickness and about 1-2 years in age is selected and thorns and side branches are removed upto a height of 20-25 cm from the ground surface. A spot is selected at a height of about 15-20 cm from soil surface in between two internodes. A horizontal cut of about 1.25 cm in length and deep enough to cut the bark only is made with the help of a sharp knife. A vertical cut is made starting from the centre of the horizontal cut about 2.5 cm long making the shape of english letter "T". The flaps are opened with the help of a knife blade or the back of the budding knife. In case of patch budding, a rectangular patch of bark of 1.25 cm × 2.5 cm is removed from the rootstock at a height of 15-20 cm from the soil surface and replaced by a similar size of bark of the selected scion having a bud. In case of "T" budding a shield shaped bud of about 2.5 cm length is removed from the scion and is inserted in the T shaped cut on the stock. Keeping the eye of the bud facing upward. Tying may be started from any end leaving about 0.6 cm above the leaf petiole for the growth of the bud. For tyding plastic film strips of 200 gauge are found as a better material. When the bud starts sprouting, the rootstock portion above the bud may be removed to boost the growth of newly grafted bud. In fruit plants where scion bark is not easily separable with wood a piece of bark containing a bud and some portion of wood may be removed from the scion and placed on a stock after giving a similar size of cut. Tyding is done as in case of shield or patch budding.

**Observation Blank**

| *Scion Used* | *Rootstock Used* | *Date of Budding* | *Number of Plants Budded* | *Time Required for Sprouting* | *Success (per cent)* |
|---|---|---|---|---|---|

**Assignments**

1. Try budding at fortnightly intervals in different varieties of citrus and find out the most suitable time for budding.
2. Try shield budding in citrus, patch budding in guava and chip budding in grapes and find out the success in each method.

# Exercise No. 14

## Objective

Layout of an orchard.

## Introduction

The fruit plants can be planted in the selected fields by any of the prevalent methods like square, rectangular, quincunx, triangular, hexagonal, contour etc. Selection of a particular system of planting depends on the factors like maximum number of plants to be accommodated per unit area, sufficient space for the proper development of each tree and orchard operation(s) to be adopted. The square and rectangular systems of planting are commonly adopted as these systems are easy to layout, allow sufficient inter space for orchard operations and accommodate maximum number of trees per hectare.

## Materials

Spade, Khurpi, Planting board, Rope, Wooden poles and pegs, Measuring tape, Chains.

## Procedure

After selecting the planting system, field is pepared and leveled properly. It is a normal practice to layout the orchard by first placing pegs where trees are to be planted. The first row is kept parallel to the edge (boundary) of the field and half as far from as the distance between the rows of the trees. Pegs are placed at the desired distance along this line (base line), beginning at half the distance from the edge of the field. In all the systems except the hexagonal, it is desirable to lay out lines at either end of the field, perpendicular to the base line and on these places the pegs are placed at the desired distances. By placing poles at the marks on the four edges of the field, the other pegs can be fixed by sigting across both ways or this can be done with the help of a rope. In laying out a field by the hexagonal system, two chains or ropes should be used each the length of the distance between trees. Both the chains are fastened to one end to a common ring and each has a ring at the other end. The rings are just large enough to slip easily over the pegs used to mark the places where the trees are to be set. After laying out the first row, the two free rings are placed over adjacent pegs and the chains are drawn, tightly and a peg driven through the common ring. This is repeated until the entire field is laid out. It is desirable to test the accuracy of the work by sighting along the lines in three directions, to see that they are straight.

**Assignments**

1. Work out the number of plants per unit area on a given spacing for different systems of planting.

2. Practice the laout of an orchard with various systems of planting particularly square, rectangular and hexagonal systems.

# Exercise No. 15

### Objective

Use of planting board for tree planting.

### Introduction

The plants occupy the same position where the pegs were placed in the field at the time of planting the tree. The peg is removed first followed by pit making to receive the plant. The plant may be with the ball of earth or without ball of earth. There is every possibility that plant may not occupy the same position where the pegs were originally placed. In order to overcome this problem, planting board is used. Use of a planting board ensures the planting of the plant in the same position.

### Materials and Equipments

Spade, Khurpi, Planting board, Wooden pegs, Rope and Chain.

### Procedure

The planting plan has been selected. Land is prepared and stakes have been fixed at the correct positions. Remove the pegs and dig the pit for receiving the ball of earth. While planting the plant in the pit, the trunk of the plant should occupy the position of pegs. Plant planted with the help of planting board ensures planting of the plant in the correct position. Place the planting board in such a way that the centre notch of the board is in the position of marking peg. Drive the pegs at each end of the board. Remove the planting board and the marking peg. Dig the pit with the help of Khurpi large enough to receive the ball of earth, where the marking peg stood. Put the tree with ball of earth in the peg. Bring back the planting board in such a manner that the two pegs (guide pegs) at both ends fit in the notches of the board. With the help of these guide pegs adjust the position of the tree. The tree should

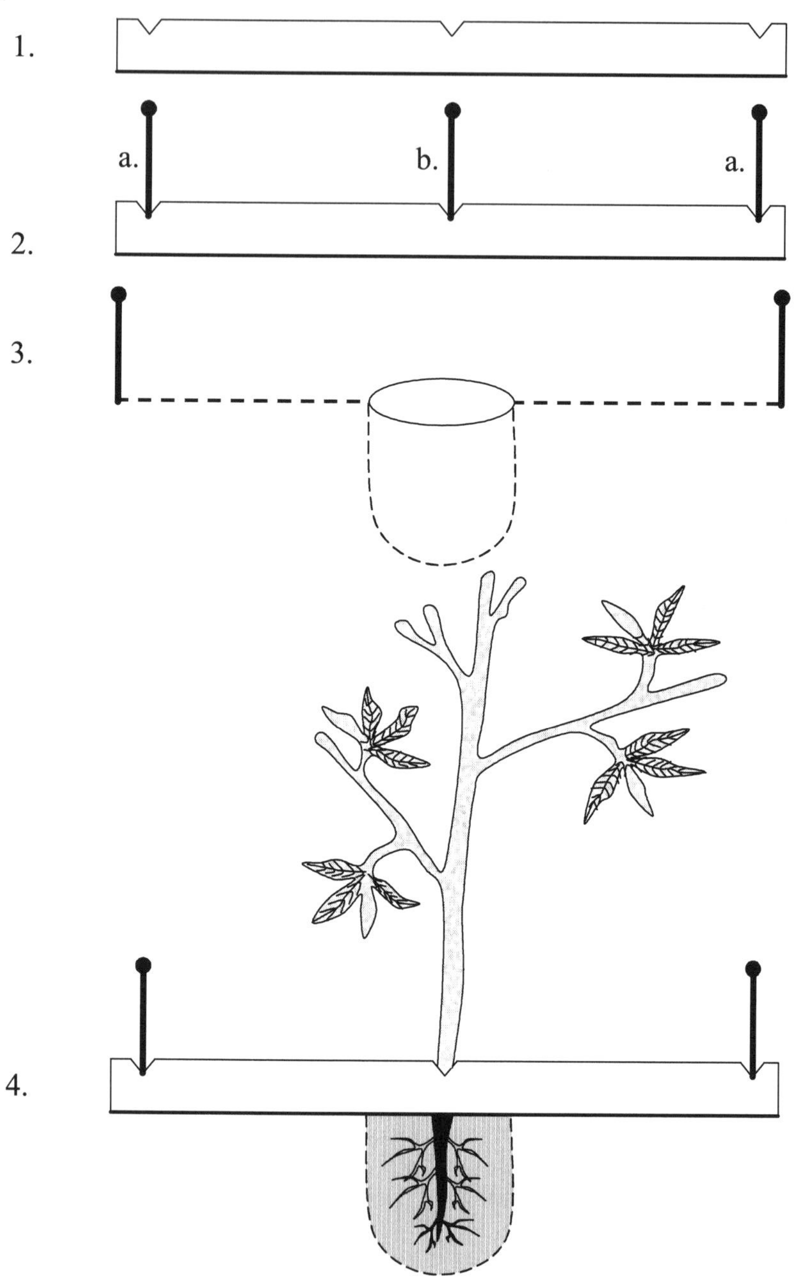


1. Planting board.
2. Planting board with marking and guidepegs.
   (a) Guide peg. (b) Marking peg.
3. Pit dug after removing the marking pegs and planting board.
4. Planted plant adjusted by bringing back the planting board.

occupy the same position which was occupied by the marking peg. After adjusting the position of the tree, adjust the depth of the plant. The plant should be planted at the same depth as they stood in the nursery. Throw loose soil all around the tree. Press the soil firmly with the help of Khurpi handle. The pit should be filled up to the ground level. Make a small basin around the tree and irrigate the plant. All the new growth and surplus growth at the time of the planting should be removed. Planting should be done either in the evenings or on a cloudy day.

**Assignments**

1. What is planting board?.
2. When planting should be done?

# Exercise No. 16

## Objective

Planting of an orchard.

## Introduction

The planting of an orchard involves a number of operations like digging of proper size of pits at the correct position, filling of pits, lifting of plants with or without earthball from the nursery and planting in the pits at the permanent site. These operations are quite technical and involve lot of skills. Slight mistake at this stage may cause the poor survival of planted trees and may affect the initial health of the orchard. Therefore, proper care must be taken while performing the planting operations.

## Materials

Spade, Khurpi, Planting board, Rope, Wooden poles, Pegs, Measuring tape.

## Procedure

For planting orchard, a pit is dug first for receiving the earthball after removing the peg. The stem of the plant should occupy the position of the peg at the time of planting. Planting board is used to ensure the planting of the plant in the correct position. Planting board is fixed in such a manner that the central notch of the board remains in the position of marking peg. Pegs are also inserted in the side notches of both ends of the board. Now planting board as well as the marking peg are removed. A hole is dug with the help of Khurpi large enough to receive the ball of earth where tree marking peg stood. Plant with the earth ball is placed in the pit. Planting board is brought back in such a way that the two guide (side) pegs at both ends fit with the notches of the board. With the help of these guide pegs, position of the plant is adjusted. The plant should occupy the same position which marking peg had occupied. After locating the site of the plant, the depth of the plant is adjusted. Now loose soil is put all around the earthball and pressed firmly with the handle of the Khurpi. The pit should be filled upto the ground level. All the new and excess growth are removed at the time of planting and this operation should be performed either in the evening or on a cloudy day. The plants are also planted bareroot during dormant season particularly in case of deciduous tree species. While planting bareroot plants, the lateral roots should be pruned back and the root portionis set in the middle of the pit with the help of loose middle of the pit with the help of loose soil. After planting, a small basin is made around the plant and irrigation is given in this basin just after planting is over.

## Lifting of Plant

Removal of plants from nursery bed with or without earthball is known as lifting of plants. Evergreen plants are commonly lifted with earthball while the deciduous plants can be lifted bareroot when dormant. The weeds around the plant intended to be dug are removed before digging. The ball size is marked with the help of a Khurpi. The earthball size will depend on the age, height and transportation distance. Seedlings and grafted plants of 1-2 years of age require an earthball size of 20-25 cm. While lifting, the soil of the plant is removed from one side upto a depth of 20-30 cm and then from remaining sides. The earthball is taken out slowly from the soil. It should be cylindrical and slightly topering towards the base. Irrigiation of the nurserybed is performed 3-4 days before lifting if the nursery bed is dry in order to ease the operation.

## Assignments

1. Practice the planting of fruit plants with and without earthball.
2. Lift two plants of mango and note the time required for lifting of each plant.
3. List factors determining the size of the earthball.

# Exercise No. 17

## Objective

Training of fruit trees.

## Introduction

Training is defined as the removal of certain parts of the plant in order to give them a proper shape and form and to develop a strong framework in early years of planting so that plants can hold the heavy load of quality crops. Training of the plants is started from the nursery itself and completed within 4-5 years after planting.

## Materials

Secateur, Pruning shear, Pruning saw, Pruning knife, Lopper, Stair etc.

## Procedure

There are three basic forms of training i.e., central leader, open centre or vase system and modified central leader system. In central leader system, the trunk is encouraged to form a central axis with branches distributed laterally up and down around the stem. The central axis or leader is the dominant and the main direction of the growth is upward. In open centre system, the main stem is terminated and growth is forced through 3-4 selected branches originating at appropriate distance on the trunk. In modified central leader system, the main leader is allowed to grow for 4-5 years and 3-4 lateral branches are selected on main trunk alternatively at 15-20 cm distance from each other. After 4-5 years, the main leader is cut back leaving 120-150 cm portion from the ground level.

## Assignment

1. Practice the various methods of training in the nursery and in the young orchards.

# Exercise No. 18

## Objective

Pruning of fruit trees.

## Introduction

Pruning means removal of certain parts of the tree in order to modify and utilize its natural habits, so that more and better fruits can be obtained at a lesser cost over a longer period of time. Pruning is achieved through various means like heading back, thinning, pinching, tip pruning, shearing, stubbing and disbudding. Depending upon this severity, the pruning may be classified as heavy pruning or light pruning. Light pruning is performed in almost all types of species and consists of removal of undesired shoots from main trunk and framework branches and removing the water sprouts, dead, diseased and insect infected shoots. Heavy pruning is done in case of trees which bear on one year old growth or on the current season growth. Thus severity of pruning depends upon the bearing habit of the tree species. Pruning operations are performed annually in order to get better size and quality of the fruits.

## Materials

Secateur, Pruning shear, Pruning saw, Pruning knife, Lopper, Pole pruner, Stair etc.

## Procedure

A person involved in purning must know the bearing habit of the tree species and its history of pruning schedule. The person should move closer to the tree and pruning should be startd from the trunk moving towards the top of the tree. First of all, the undesirable shoots arising from the trunk should be removed totally from their base. Root suckers may also be removed. The water sprouts, dead, diseased and the branches infested with insect-pests may also be removed. The undesirable branches cris crossing each other may be thinned making the centre open. Now depending upon the bearing habit of the fruit tree species, decisions are made accordingly. The fruit species which bear on currect seasons growth or one year old shoots are pruned heavily every year or alternate years. The tree species which bears on spurs may be pruned depending upon the life of the spurs. Spurs of certain plant species are long lived while other are short lived. The pruning may be done in such a way that sufficient bearing spurs may be available every year to produce an economical crop. The old unproductive spurs are removed every year and the new spurs are encouraged in order to make the balance between bearing and non-bearing spurs. Usually, the pruning is done in dormant season which is called as dormant pruning. In some species, an additional pruning is done during the summer which is called as summer pruning and is performed to avoid too much growth, however, it should be done in exceptional cases.

**Observation Blank**

| *Sl.No.* | *Name of Crop* | *Intensity of Pruning* | *Yield per Tree* | *Fruit Size* | |
|---|---|---|---|---|---|
| | | | | *Length* | *Diameter* |
| 1. | Ber | | | | |
| 2. | Peach/Pear | | | | |

**Assignments**

1. Practice the pruning in Ber, pear and peach.
2. Perform the various operations of pruning like thinning, heading back, pinching and list the tools which are to be used in performing such operations.

# Exercise No. 19

## Objective

Yield estimate of an orchard.

## Introduction

Annual yield estimate of the orchard provides useful informations to the orchardist. It shows the increase or decrease in the yield over last year. Information such as response of trees to the kind and or dose of fertilizers and of cultural practices. The extent of loss due to certain pest and diseases can be known. This study is very useful for selecting mother trees.

## Materials

Basket and Metal tags.

## Procedure

Prepare a map of the orchard and select ten healthy, uniform trees randomly at flowering for the purpose of yield estimate. Mark the selectd trees in the map. Tag with number on each selected trees. Harvest the ripe fruits of the selected trees daily or as necessary. Sort the fruits based on the size and grade them. Record the weight and number of sound, pest and diseased fruits of each tree separately. Record the observations as given below. With the help of observation, the informations required, can be worked out.

**Observation Blank**

| *Number of Trees Studied* | *Month and Date of Harvesting* | *Number of Fruits* | | | *Weight of Fruit (kg)* | | | *Pest, Diseased and Culled Fruits* | | *Percentage* | | |
|---|---|---|---|---|---|---|---|---|---|---|---|---|
| | | *L* | *M* | *S* | *L* | *M* | *S* | *Weight kg* | *Number* | *L* | *M* | *S* |
| 1. | | | | | | | | | | | | |
| 2. | | | | | | | | | | | | |
| 3. | | | | | | | | | | | | |
| 4. | | | | | | | | | | | | |
| 5. | | | | | | | | | | | | |
| 6. | | | | | | | | | | | | |
| 7. | | | | | | | | | | | | |
| 8. | | | | | | | | | | | | |
| 9. | | | | | | | | | | | | |
| 10. | | | | | | | | | | | | |
| Total: | | | | | | | | | | | | |

**Average**

| *Tree Number* | *Large* | | *Midium* | | *Small* | | *Culled* | | *Total (No.)* | | *Weight (kg)* | |
|---|---|---|---|---|---|---|---|---|---|---|---|---|
| | *No.* | *Wt.* | *No.* | *Wt.* | *No.* | *Wt.* | *No.* | *Wt.* | *Culled* | *Sound* | *Culled* | *Sound* |
| 1. | | | | | | | | | | | | |
| 2. | | | | | | | | | | | | |
| 3. | | | | | | | | | | | | |
| 4. | | | | | | | | | | | | |
| 5. | | | | | | | | | | | | |
| 6. | | | | | | | | | | | | |
| 7. | | | | | | | | | | | | |
| 8. | | | | | | | | | | | | |
| 9. | | | | | | | | | | | | |
| 10. | | | | | | | | | | | | |
| Total: | | | | | | | | | | | | |

1. $$\text{Average yield per tree} = \frac{\text{Total yield of selected tree}}{\text{No. of trees selected}}$$

2. Yield of the orchard = Average yield per trees × No. of trees in the orchard

3. $$\text{Percentage of culled fruits} = \frac{\text{No. of culled fruits per tree} \times 100}{\text{No. of sound fruits per tree}}$$

**Assignments**

1. Calculate the yield estimate of an orchard of mango cls. Dashehari and Chausa of 64 years old.

2. Find out the yield estimate of 20 years old orchards of orange cl. Mosambi and compare it with cultivar Sathgudi.

# Exercise No. 20

**Objective**

Preparation of seed bed for vegetables and sowing of seeds.

**Introduction**

Seed for vegetable nursery should be prepared in a proper way to provide an ideal condition for seed germination and subsequent seedling growth. There condtions may be the pulverized soil having sufficient amount of organic matter, free from insect-pests, nematodes and other pathogens which can hold proper amount of moisture with proper drainage. Such ideal conditions are never found under natural conditions, and therefore, soil manuring and treatment are needed. Usually sandy loam soil with sufficient amount of organic matter is preferred for the preparation of seeded and seed sowing.

**Materials**

Spade, Khurpi, Rope, Measuring tape, Hand dibbler, Hand cultivator, Hand fork, Seeds, Insecticides, Fungicides, Weedicides and Antibiotics.

**Procedure**

A good deep soil with sunny site should be selected for seedbed preparation, the soil should be dug with the help of a spade and clods are broken with the help of the back of the spade or Khurpi in order to make the soil loose and fine. The soil is raked vigorously to remove weeds. If the soil selected for nurserybed preparation was previously occupied with forest plantation, it should be sterilized through steam or with farmaldehide to kill the spores of pathogens. After treating the soil, it should be covered with polyethylene sheets for 7-10 days. Soil solarization may also be done after covering the soil with polyethylene sheets and sealing its sides and leaving as such for two months during hot summer days. In this way, soil is fully sterilized without giving any injurious effect. After soil treatment, it is again dug and leveled. The channels for drainage and irrigation may be prepared. Then, the soil is raised 10-15 cm high and 1.25 metres wide. The length of the bed depends upon the convenience. A strip of 60 cm is left between two beds for walking and this space can be utilized for drainage and irrigation. A well rotten FYM @ 10-15 kg/m$^2$ should be mixed in the upper layer of bed and leveling is done in such a away that the middle portion of the bed remain slightly higher than both sides in order to drain the water. Now these beds are ready for sowing. One of the problems in sowing of seeds is to find out the rate of sowing to get desired stand of seedlings. The final stand of the seedlings is determined in order to produce the maximum number and high quality seedlings. Once the desired final stand is determined, the required rate of sowing can be calculated, if the germination percentage, purity percentage and the number of seeds per kg are known.

It can be calculated by the following formula:

$$\text{Kilogram of seeds require/unit area} = \frac{\text{No. of seedling required/unit area}}{\text{No. of seeds/kg} \times \text{Germination per cent} \times \text{Prurity per cent}}$$

Once the amount of seed is calculated, the beds are marked with lines at a distance of 5-10 cm in which seeds are sown 1-2 cm deep. Actually the depth of sowing depends upon the size of seeds, the condition of seedbed and the season of sowing. With too deep sowing, germination may fail because the seedlings are not able to emerge before the food reserves in the seed are exhausted. In addition the poor aeration may occur at lower depth. The proper depth is also related to water supply. Too shallow planting will place the seeds in the top layers which are subjected to drying. A general rule is to sow the seeds 3-4 times their thickness. Before sowing, the seeds are treated with fungicide in a ratio of 1:500 (Thiram 2 g/kg of seeds) to check the damping off. After placing the seeds at proper depth in the lines, these are covered with soil : FYM mixture (1:1) and the beds are covered with dried grasses or polyethylene films just to protect the seeds from adverse environmental conditions. As soon as seeds start germinating, the cover should be removed.

**Observation Blank**

| *Crop* | *Amount of Seeds Sown in the Nursery Bed* | *Germination (per cent)* | *Number of Seedlings Ready for Transplanting* |
|---|---|---|---|
| Tomato | | | |
| Brinjal | | | |
| Chilli | | | |
| Cauliflower | | | |

## Assignments

1. Prepare a nurserybed for sowing of brinjal seeds.
2. Calculate the amount of seeds required for sowing per unit area of vegetable nurserybed.

# Exercise No. 21

## Objective

Care of vegetable seedlings in the nursery and their transplanting.

## Introduction

care of vegetable seedlings in the nursery is a must as they grow faster and take lesser time to be ready for final transplanting in the field. Such cares include regular watering of seedlings, protecting them from hot scorching sun, diseases, insect-pests and weeds at a proper time without affecting their growth. Regular observation is required to save the plants from various biotic and abiotic stresses. Transplanting refers the process of lifting of seedlings or plants from one environment to another. In case of vegetable seedlings, shifting is done from seedbed to their final growing site in the field in order to provide them better space for their proper growth.

## Materials

Spade, Khurpi, Rope, Measuring tape, Sprayer, Insecticides, fungicides, acricides, weedicides and antibiotics, Watering can.

## Procedure

After germination of the seeds, frequent light irrigation is needed in the nursery. Regular weeding and hoeing operation are performed. If there is attack of insect-pests and diseases, proper sprayings are done at proper time. Usually 1-2 months old seedlings are used for transplanting. The seedlings are required to be hardened by reducing. The interval of irrigation atleast 15 days before transplanting. For transplanting field prepared by one to two ploughing followed 2-3 harrowings and planking, then leveling is done keeping the drainage and irrigation points in mind. Manures and half of the nitrogenous and full amount of phosphatic and potassic fertilizers are mixed uniformly in the soil based on the requirement of the crop. Lines are, then drawn at a proper distance which varies with crop to crop. Seedlings from the nursery are uprooted with the help of the Khurpi without damaging the root. The weaker and injured seedlings are discarded at this state. These seedlings are kept in the bamboo baskets and are carried to the transplanting field. They are kept in shady place till they are transplanted. It is better if the seedbeds are irrigated atleast 48 hours before uprooting of seedlings to facilitate uprooting and to keep seedlings in a fresh condition. Method of transplanting varies with the size of operation. If the transplanting operation is larger in size, one can go for mechanical transplanting adjusting plant to plant and row to row distance in mechanical transplanter. If it is done on smaller scale, the human labourer are employed to transplant the seedlings at a proper distance within the row drawn previously at a proper distance with the help of a rope. Immediately after transplanting, a light irrigation is done. If there is shortage of irrigation water, individual transplant may be irrigated with the help of a can.

## Observation Blank

| *Crop* | *Survival of Transplanted Seedlings in the Field (per cent)* |
|---|---|
| Brinjal | |
| Tomato | |
| Cauliflower | |

## Assignments

1. Transplant the brinjal, tomato and cauliflower seedlings in the field.

# Exercise No. 22

## Objective

Identification of ornamental Plants.

## Introduction

A number of plant species are grown for various purposes in the garden which belong to a number of families and ranged from small ferns to the gigantic trees. Exposure of students to various ornamental plants is a first lesson in teaching of ornamental horticulture. The student should be well acquaint with the common english names, scientific names and families of various ornamental plants. Besides, they should be able to identify the various cultivars of a species. The identification of plant species or cultivar is generally done by observing their growth habit and morphological characters such as colour, size, nature of leaf, flower, fruits and seeds. Identification of plants through morphological characters is a simple and cheapest way but for a beginner, it is a difficult task. This difficulty can be solved as soon as they develop their interest and come closer to the plants. More closer their involvement, more easier it is to identify. It is better to start with a few plants. Frequent visits of the garden in the leisure hours make identification more easier but in the beginning the visit should be alongwith the expert who may explain the major differences between two cultivars or species.

## Materials

Pencil, Notebook, Scissors, Blotting paper, Paper bags, Label.

## Methods

The first step in identification is to arrange a layout plan of the garden. Compare the plants with the layout plan. Label them properly. Select few cultivars of a species or few species of a family and study their morphological characteristics and differences between them. Note down the specific characters in the note book. Few leaves or flowers may also be collected with the proper label and pressed between two blotting papers till they dry. These specimen parts may be pasted on the sheet of a thick paper with the help of a cellophane tape and are labeled properly. For identification of a large number of plant species/cultivars, a regular practice is required.

## Observation Blank

| *Sl.No.* | *Common or English Name* | *Botanical Name* | *Family* |
|---|---|---|---|
| **A.** | **Annuals** | | |
| a. | Summer | | |
| 1. | | | |
| 2. | | | |
| 3. | | | |
| b. | Winter | | |
| 1. | | | |
| 2. | | | |
| 3. | | | |
| **B.** | **Biennial** | | |
| 1. | | | |
| 2. | | | |
| 3. | | | |
| **C.** | **Shrubs** | | |
| i) | Flowering shrubs | | |
| 1. | | | |
| 2. | | | |
| 3. | | | |
| ii) | Foliage shrubs | | |
| 1. | | | |
| 2. | | | |
| 3. | | | |
| iii) | Hedge plants | | |
| 1. | | | |
| 2. | | | |
| 3. | | | |

| Sl.No. | Common or English Name | Botanical Name | Family |
|---|---|---|---|
| iv) | Edge plants | | |
| 1. | | | |
| 2. | | | |
| 3. | | | |
| **D.** | **Trees** | | |
| i) | Flowering | | |
| 1. | | | |
| 2. | | | |
| 3. | | | |
| ii) | Avenue | | |
| 1. | | | |
| 2. | | | |
| 3. | | | |
| **E.** | **Cacti and Succulents** | | |
| i) | Cacti | | |
| 1. | | | |
| 2. | | | |
| 3. | | | |
| ii) | Succulents | | |
| 1. | | | |
| 2. | | | |
| 3. | | | |
| **F.** | **Ferns** | | |
| 1. | | | |
| 2. | | | |
| 3. | | | |
| **G.** | **Palms and Cycades** | | |
| i) | Palms | | |
| 1. | | | |
| 2. | | | |
| 3. | | | |

| Sl.No. | Common or English Name | Botanical Name | Family |
|---|---|---|---|
| ii) | Cycades | | |
| 1. | | | |
| 2. | | | |
| 3. | | | |
| **H.** | **Orchids** | | |
| 1. | | | |
| 2. | | | |
| 3. | | | |
| **I.** | **Lawn and Turf grasses** | | |
| i) | Lawn grasses | | |
| 1. | | | |
| 2. | | | |
| 3. | | | |
| ii) | Turf grasses | | |
| 1. | | | |
| 2. | | | |
| 3. | | | |
| **J.** | **Water and Marshyland Plants** | | |
| i) | Water plants | | |
| 1. | | | |
| 2. | | | |
| 3. | | | |
| ii) | Marshyland plants | | |
| 1. | | | |
| 2. | | | |
| 3. | | | |
| **K.** | **Mangroves** | | |
| 1. | | | |
| 2. | | | |
| 3. | | | |

**Assignments**

1. Prepare a list of plants giving their common english names, botanical names and families.

2. Classify the plant species according to their growth habit, time of flowering and colour of flowers.

# Exercise No. 23

## Objective

Seedbed preparation and sowing of ornamental plants.

## Introduction

Seeds of ornamental plants are usually sown in the flate or raised seedbeds in the nursery. Therefore, the seedbed preparation is an important aspect of seedling raising because the germination of seeds and subsequent growth of seedlings are influenced by the seedbed. Nursery production requires a fertile, well drained soil of medium to light texture. Preplant fumigation and weed control are essential aspects of most nursery operations. Selection of flate or raised seedbed depends upon the season. In dry growing season flate beds are prepared while during wet season the beds are raised in order to provide proper drainage and aeration.

## Materials

Rope, Spade, Khurpi, Well rotten F.Y.M. or compost, Seeds, Fungicides like thiram or cerasan or Agrosan G.N., Mulching material (Dried weeds or leaves), Spirit level.

## Methods

Well drained fertile medium textured soil is selected for seedbeds. The selected soil should be ploughed deep enough in the previous summer and left as such for one or two months or it may be harrowed and planked to make it fine and the beds are prepared which are covered with white polyethylene sheet sealing its ends to achieve solarization. In solarization process soil becomes sterilized without loosing useful microorganisms or organic matter. These beds are solarized during the summer and left as such till final seedbed preparation. During seedbed preparation, the plastic sheets are removed, soil is again ploughed and harrowed to the fine texture. The plot is then leveled and drains are provided. A common size of seed bed is 1.2 m wide with the length varying according to the size of the operation. Beds may be raised to 30-45 cm to ensure good drainage. Beds are separated by walkways of 60 cm wide. The beds should be faced with north-south orientation as it gives more even exposure to light than east-west. The middle of the bed should be slightly raised than the sides. Seeds should be treated with the suitable fungicides at the rate 2 grams per kg of seeds. These treated seeds may be either broadcasted over the surface of the bed or drilled into closely spaced rows (5-10 cm) with seed planters. For economy, seeds should be planted as closely together as feasible without over crowding in order to avoid damping off and to increase vigour and size of the seedlings. Dense sowing results thin, spindly plants and small root systems which do not transplant well. Depth of planting varies with the kind and size of the seeds. In general a depth of 3-4 times, the diameter of the seed is satisfactory. Seeds can be covered by a sterilized mixture of soil, coarse sand and well rotton F.Y.M. (1:1:1). The beds are finally covered by various mulches like dried grasses, polyethylene sheet (in winter) etc. Once the seeds start coming out of the soil, the mulches must be removed. The seeds may be planted any time of the year (Spring, Summer, Rainy season) depending on the dormancy conditions of the seed, the temperature, the management practices and the location of the nursery. The optimum seed density depends on the species and on the nursery objectives. Once the actual density is determined, the necessary rate of sowing can be calculated from the data obtained from germination test. The following formula is useful in calculating the seed rate for sowing.

$$\text{Weight of seeds to Sow/unit area} = \frac{\text{Density (Plants/unit area) desired}}{\text{Purity (per cent)} \times \text{Germination (per cent)} \times \text{Field factor*} \times \text{Number of seeds/unit weight (Seed count)}}$$

* Field factor is a correction term applied based on the expected losses which are experienced at a particular nursery with a particular species. It is a percentage expressed as a decimal.

**Observation Blank**

| *Sl.No.* | *Species/Cultivar* | *Number of Seeds Sown* | *Number of Seeds Germinated* | *Germination (per cent)* |
|---|---|---|---|---|
| | | | | |

**Assignments**

1. Prepare a seedbed for sowing of winter season annuals.
2. Calculate the seed rate by using above formula.

# Exercise No. 24

## Objective

Potting and repotting of ornamental plants.

## Introduction

Potting refers to planting in pots containing soil mixture while potting-on refers the transferring of plant from a smaller to a larger pot with a fresh soil mixture. When the plants grow to a sufficient height and the pot size is not adequate, the plants should be transferred to the next larger size pot as the original pot gets filled with roots when the plant grows. Repotting is generally referred to the transfer of a plant from one pot to another replacing the soil mixture with the fresh one. In India, the burnt porous clay pots of various designs and sizes are used for various purposes in ornamental horticulture. The chief uses of pots are in propagation of plants, cultivation of plants and flowering annuals. Now a days, plastic pots and tubes are becoming popular. Besides flates, fiber pots, paper pots, wood and metal containers are also used for various purposes. The best time for potting and shifting of plants is rainy season.

## Materials

Pots, Crocks, Various components of potting mixture, Khurpi, Watering can, Trowel.

## Method

### A. Potting Mixture and Media

Potting mixture and media consist of a mixture of organic and inorganic components. The organic component includes: peat, softwood and hardwood barks, sphagnum moss, saw dust, leaf mulch and rice hulls. A coarse mineral component is used to improve drainage and aeration by increasing the proportion of large, airfilled pores. A variety of mineral components include: sand, grit, pumice, perlite, vermiculite, clay granules and rockwool. There is no single ideal mixture. An appropriate medium depends upon the species, season, purpose, cost and availability. A good medium must be sufficiently firm and dense with constant volume while wetting and drying. It should be well decomposed, sufficiently porous, free from weed seeds, nematodes and various pathogens and having good waterholding capacity. It should have a high cation exchange capacity and readily available at acceptable cost. The growing media for the pot culture is given in Table 1.

**Table 1: Potting Mixture and Growing Media**

| | *Composition (In parts V/V)* | | | | | | | |
|---|---|---|---|---|---|---|---|---|
| *Potted Plant* | *F.Y.M. Compost* | *Leaf Mould* | *River Sand* | *Red Soil* | *Bone-meal* | *Brick Powder* | *Lime Powder* | *Char-coal* |
| Anthuriums and Phelodendrons | 2 | 4 | 2 | – | 1 | 1 | – | 1 |
| Begonias and Caladiums | 4 | 4 | 2 | – | 1 | – | – | ½ |
| Cacti and Succulents | 2 | 2 | 2 | – | ¼ | 2 | – | – |
| Bulbs and tubers | 2 | 1 | 2 | 1 | 1 | – | – | – |
| Ferns and Palms | 2 | 4 | 1 | – | ½ | – | – | – |
| Roses | 5 | – | 2 | 3 | 2 | – | – | – |
| For all plants | 3 | 2 | 2 | 1 | ½ | – | – | – |

* A small quantity is needed to get slight alkaline soil.

## B. Filling of Pots

Select well baked and sound flower pots of the proper size. The size of the pot will depend on age of the plant and the purpose for which pots are filled. Wash the pots both inside and outside with clean water. Place a large crock on the drainage hole. Now put several smaller crocks to a depth of 5-10 cm, depending upon the size of the pot. Put 1.25 to 2.5 cm layer of coarse sand or coconut fibre. This is necessary to check away the washing out of fine soil particles. Fill the pot with the soil mixture or compost. When half full, press the mixture firmly. Continue filling upto the rim of the pot. Press the potting mixture again and finally fill and press mixture to a point where a room of 2-3 cm is left for holding water.

## C. Potting

The filled pots can be used for sowing of seeds, potting of plants or planting cuttings. For planting of pots a hole is scooped out in the centre of the pot. The plant is kept in the centre of the hole with roots well distributed in all directions. The soil mixture is placed all round the plant and pressed firmly and uniformly. Care must be taken that planting is not done too deep. The plant is irrigated with watering can fitted with a fine rose. The potted plant is then placed in a cool shady place.

## D. Repotting

The first step in repotting is the depotting. A simple technique is to be followed for removing the plant intact from the pot. The pot is lifted by right hand, the palm spreadover the top of the soil holding the steam between second and third finger and the thumb along the side of the pot. The pot is then turned upside down. If necessary a gentle tap is given on the rim of the inverted pot against any solid base or on the edge of the bench to loosen the earth ball. The whole earth ball with its enterwining roots will slipout as one piece. If the soil surface is dry, the surface soil around the plant close to the rim of the pot is removed using a trowel to make the plant along with the soil come out easily. The ball of earth then taken out from the pot carefully. Before putting the plant in the new pots the lower finer roots and some soil are removed. If necessary the pot are irrigated lightly atleast 3-4 hours before shifting. The plant is placed in the centre of the new pot and then the sides are packed with the soil mixture use. A little moist compost is added before passing down the roots lightly. Periodical repotting should be carried out for perennials as the overgrown root system produces pressure against the sides of the pot and further growth of root system is affected. Repotting either in the same pot or in slightly larger pot is preferred after pruning the old and dead roots to provide more space for new roots. Pruning the top of the plant at this stage is necessary reduce transpiration.

### Assignments

1. Fill the pots for planting of cacti and succulents.
2. Shift the potted rose plant in a larger pot.
3. Prepare the potting mixture for planting of bulbs and tubers.
4. Write the characteristics of a good potting mixture.

# Exercise No. 25

## Objective

Training and pruning of ornamental plants.

## Introduction

Training of ornamental plants refers the removal of certain parts of the plants in order to give them certain shapes. A few perennial herbs such as begonias, geraniums and foliage plants like pothas, philodendron are amenable to shaping by pruning and clipping of the young branches. Frequent training and shaping of the hedge and edge plants are essential to keep them compact and elegant looking. Pruning and clipping of small trees and woody shrubs in fanciful or obstract shapes are being practiced from ancient time and is reffered as 'Topiary'. The shrubs and trees are used to shape into domes, cones, spheres, umbrellas, birds, animals etc. The climbers are trained as standard arches or pergolas. Certain plants are trained as bonsai. Pruning refers the removal of certain undesirable parts of ornamental plants in order to maintain them into desirable health so that they may bear quality foliage or more quantity and quality flowers for longer period of time. It may be the removal of dead, diseased, insect infected shoots, water sprouts, suckers, thinning of foliage and flower buds (disbudding) and removal of abscised flowers and unripened fruits (defruiting in still flowering plants). Some times root pruning is practiced in certain plant species although it is not a desirable practice.

## Materials

Knives, Secateur, Pruning shears, Pruning saw, Different types of plant material such as annuals, biennials, climbers, woody shrubs, trees etc.

## Method

### Training

The training of ornamental plants depend upon the nature of the species and the purpose for it is grown. The hedges and edges are trained in certain shapes without any support whereas topiaries are trained initially on certain types of structures made up of aluminium or steel wires or wooden sticks. Once the plants are planted in the base of the structure and grown few centimeters above the structure, they are clipped to the shape of desired structure which may be bird or animal or human being. The potted/bedded flowering plants such as chrysanthemum, Dahlia etc. are trained either on single or multistem plants depending upon the choice of the grower and purpose for its grown. Such plants are maintained by giving a support or by staking.

### Pruning

All such plants which are trained as hedges, edges, topiaries etc. are maintained in certain shapes by clipping of new growth regularly and are maintained on single or double stem by removing all undesirable growth like suckers, water sprouts, dead, diseased and insect-pests infected branches or foliage. After pruning of such plants protective sprays against pathogenic organisms or insect-pests are desirable so that the pathogens may not enter from cut ends. In flowering plants, all dried flowers are removed regularly so that the new growth may initiate and develop into new flower crop as in case of roses and bougainvilleas. In case of potted plants disbudding operation is performed. If a single big size flower is needed for exhibition, all lateral flower buds are removed. (pinched). If one wants to get more number of small flowers per plant, or pot, apical flower buds are removed. In case of arches or pergolas, all the hanging undesirable shoots are removed particularly during the growing season and the shape is maintained by pruning into the desired form. The time of pruning is an important aspect and depends upon the objectives of the grower. The roses are mainly pruned in the October to get the flowers at the time of christimas. In October the roses are cutback by leaving few centimeters of branches having atleast one bud and three to four such branches are left. All the suckers arising from the rootstock should be removed. The plants should be maintained by regular removing of suckers from the rootstocks. In roses, root pruning is also practiced although it is not desirable for the plant growth. The pruning intensity in roses depends upon the varieties. The hybridtea varieties withstand extensive pruning while floribundas require light pruning. The miniature ones need removal of the dead and weak shoots and clipping of the top. In case of standard trees or shrubs or climbers pruning is performed during winter season or when the buds are in resting stage. At this period all the undesirable, dead, diseased and insect-pests infected branches are removed once in a year. Another pruning of such plants should be done just after flowering to remove dried flowers. The fruits should also be removed if seed collection is not an objectives. This will shave the energy of the plants and encourage growth and flowering. While pruning the plants, care should be taken that the operation should be performed with the help of the sharp tools to get the clear cuts.

## Assignments

1. Practice the pruning of roses and disbudding and pinching in potted plants of chrysanthermums and Dahlias.
2. Practice the training of a topiary in the shape of a bird. Name the plants which are to be used for topiary making.
3. Practice the clipping of a hedge and an edge.

# Exercise No. 26

**Objective**

Cultural practices in important ornamental plants.

**Introduction**

There are certain operations that are to be followed judiciously for successful cultivation of ornamental plants. Most of these operations are of vital importance for the growth and production of the plants. However, some other operations are designed to enhance the aesthetic value of the plant material. The cultural practices vary with the plant type, gardener's objective, etc. The general cultural practices which are routinely followed are being discussed here.

**A. Annual and Bulbous Plants**

Incase of annuals, the well prepared land should be incorporated with 5Kg well rotten FYM, 10 g each of N, $P_2O_5$ and $K_2O$ per square meter and irrigated lightly whereas in case of bulbous crops, 20g each of N, $P_2O_5$ and $K_2O$ should be added. Transplant the seedlings in the cool hours of evening at a spacing of 30-40 cm, 15-20 cm and 10-12 cm for tall, medium and dwarf annual plants, respectively and the beds should be irrigated immediately. Bulbs should be placed at 7-10 cm depth at a spacing of 30×20 cm. Regular weeding and hoeing operations are essential for healthy plant development. Regular irrigation as per the requirement during all stages of development is highly beneficial. Generally 7-10 days interval in winter and 5-7 days in summer is appropriate. In annuals, after one month of transplanting of seedling another dose of N (10g/m$^2$) should be broadcasted whereas in bulbous crops N @ 40g/m$^2$ into two-three split doses should be applied. Most of the bulbous crops and many of the annuals with weak, slender or straggling stem should be staked properly. The creepess like sweet pea, morning glory, nasturtium etc should also be given proper support. Pinching of certain seedlings like carnation, marigold and dianthus should be done whereas in aster, cosmos and zinnia develop flower buds as early as when they are only 5-10 cm high, should be pinched off. Disbudding of the auxillary buds should be done as soon as they appear to reduce the number of buds per stem for large size flowers. In annuals, remove the off types before flowering or flower opening. Upon maturity seeds should be collected before splitting of seedpods and treated with insecticide and packed in muslin bags with silica gel and stored best in airtight tins. In bulbous crops, after 10-12 weeks after flowering bulbs are lifted. Water should be withheld before lifting the bulbs. Air dry the lifted bulbs in shade. Treat the bulbs with 0.2 per cent bavistin solution for 30 min and thereafter store the bulbs.

### B. Shrubs and Trees

Pits of 60 $cm^3$ size for shrubs and 1.0 $m^3$ size for trees should be dug. Four to six kilograms of well rotten FYM manure may be mixed with the excavated soil before refitting. Soak the pit thoroughly with water and when soil settles down, planting should be done. The stem end near the earth ball is placed a little below the level of the ground, fill the pit and press the soil gently and water it copiously. Utmost care is needed to keep the saplings alive and healthy. A tree guard in the form of bush or bricks should be provided to protect the sapling from animals and children. Regular weeding and hoeing are necessary for proper growth of the plant. Saplings and young plants should be watered at regular interval for proper growth and development till they develop prope root system. Prun the trees and shrubs to retains good shape and promote growth of flowers besides removing congestion or dead wood in subsequent years. Evergreens are generally not pruned except for occasional shaping. In case of any mortality, the gap filling should be done immediately and with the same age of the plant.

### Observation Blank

| *Ornamental Crops* | *Cultural Practices* | | | | | |
|---|---|---|---|---|---|---|
| | *Sowing Time* | *Seed Rate* | *N P K Applied* | *Time of Application* | *Irrigation* | *Other Operations* |
| 1. Annuals | | | | | | |
| 2. Bulbous | | | | | | |
| 3. Shrubs | | | | | | |
| 4. Trees | | | | | | |

### Assignments

1. Give the cultural practices of annuals.
2. Give the sowing time and seed rates of bulbous crops.

# Exercise No. 27

## Objective

Preparation of bonsai.

## Introduction

The word "Bonsai" is a combination of two Japanese words Bon (meaning shallow pan) and sai (meaning plant), which can be translated as "tray planting". By growing the trees in shallow containers the growth is checked but the plant is not starved. The great Japanese expert on Bonsai, Kyozo Morata said "Like a pet animal, it needs water, sunshine and nutrition". Today, Bonsai is fast becoming a popular garden art. Where an ordinary plant material is transformed into a piece of living miniature sculpture, an illusion of nature in miniature form. These fascinating trees, in addition to being grown outdoors, are great favourities for growing indoors in such places as living rooms, verandas, terraces, window sills and conservatories. The purpose of Bonsai has always been the appreciation of the total appearances of the plant. The main requisites are:

1. The plant used as bonsai must have all the vitalityof its natural form.
2. The trunk should have the appearance of age, preferably characteristics of large trees that are several decades old.
3. The branches should be trained in such as way so as to give an artistic appearance.
4. There must be perfect harmony between the shape of the container and the appearance of the plant so that the overall affect will be one of stability.
5. Plants may be so chosen that they will betray only the minimum naturalness.

## Materials

Container, Soil, FYM, Leaf mould, Fertilizers (SSP, MOP, some micronutrients like copper, lime, boron etc.), Stone, Rocks, Wire, Secateur, Khurpi, Water, Plant etc.

## Method

The three elements of bonsai etc:

(1) The container in which the trees are planted (2) the soil and (3) the plant growth. The first thing is to collect the suitable type of plant. A plant for bonsai may be raised by seeds, cuttings, layering or plants collected from the wild. While selecting a container, it must be kept in mind that the container must be shallow with any shape. A container should also have one or two drainage holes. A bonsai remains in its container for atleast one year and some may remains for two to three years without repotting.

## Potting and Repotting

While potting or repotting, the normal precautions as taken for other plants are followed. The drainage holes are covered with crocks and other materials and potting media should be filled in. A good potting media will be made up of 2 parts of good loam, $1^{1/2}$ parts coarse river sand and 1 part leaf mould. The potting mixture should not be packed too hard as this impairs aeration. A straight bonsai is planted in the centre of the pot, where as cascade type should be planted on one side. After potting, the plant should be kept in shade for a few days. The repotting operation is essentially the same as for potting. The fast growing plants will need repotting every year, while the slow growing ones may be repotted once in every two or three years.

## Style

General categories that bonsai fit into are: straight trunk, slanting trunk, twisted trunk, cascade tree, semi-cascade tree, twin trunks, ground trees, combination of trees and trees planted on rocks. Decision about the style to be adopted is taken at the outset and with regard to a close examination of the overall appearance of the tree as viewed from the front.

## Wiring

With the help of a wire, the shape of the branches and trunk can be altered. The plants may be preserved in the present shape for many years or modified in a different form by altering the existing one or to create a totally new form. Copper wire of varying gauge (a thin gauge wire for a thin branch and a thicker or lower gauge wire for the trunk or a thick branch) are used.

## Pruning and Pinching

In establishing the bonsai, training and pruning are followed together. A bonsai is kept dwarf not by starving but by various methods of pruning *viz.*, shoot pinching, leaf pinching and root pruning.

**Observation Blank**

| *Sl.No.* | *Plants to be Planted* | *Time of Planting* | *Style of Bonsai* |
|---|---|---|---|
| 1. | Flowering plants | | |
| 2. | Fruit bearing plants | | |
| 3. | Conifers | | |
| 4. | Foliage plants | | |

**Assignments**

1. Prepare the bonsai using various types of plant species.
2. Fill the observation blank after consulting the literature.
3. Describe the various styles of bonsai.

# Exercise No. 28

## Objective

Planning and layout of gardens.

## Introduction

The word "garden" means an enclosed space and hence gardening distinguished from agriculture by being within enclosure instead of open fields. The planning of garden is an art. It is said that a poet is born, and in that sense a garden architect is also born. Gardening being deeply related to the human as well as to nature. Its style has been developed according to the political and geographical conditions of a particular country. All the different styles of gardening may be divided into three broad classes *i.e.*, formal, informal and wild styles of gardens.

## Materials

Drawing sheet, Pencil, Scale, Eraser, Measuring tape, Rope, Pegs, Ranging rod.

## Method

Before laying out of any style of garden in the field, first of all, its plan should be drawn on a drawing sheet. For better planning, it is necessary to know about the following three styles of gardens.

1. **Formal Gardens:** A formal garden is laid out in a symmetrical or a geometrical pattern. In such gardens, everything is planted in a straight lines. If there is a plant on the left hand side of a straight road, a similar plant must be planted at the opposite place on the right hand side. The flower beds, borders and shrubbery are arranged in geometrically designed beds.

2. **Informal Gardens:** In an informal garden, the whole design looks informal, as the plants and the features are arranged in a natural way without following any hard and fast rules but here also the work has to proceed according to a set and well thought out plan, otherwise the creation will not be attractive and artistic. The idea behind this design is to imitate nature.
3. **Wild Garden:** A comparatively recent style of gardening. 'Wild Garden' was expounded by William Robinson in the last decade of the nineteenth century. The concept of wild garden is not only against all formalism but it also breaks the rule of land scape styles. His main idea was to naturalize plants in shrubberies. He also preached that grass should remain unmoved, as in nature, and few bulbous plants should be grown scattered in the grass to imitate a wild scenery. He also suggested that passages should be opened in the wood land, and trees, shrubs and bulbous plants should be planted among the forest flora to fulfil his idea of a wild garden. His other idea was to allow the creepers to grow over the trees naturally imitating those of the forests.

## Assignments

1. Draw a plan of layout of various styles of gardens on a drawing sheet showing their various components.
2. List out various plant species used in various types of gardens under Indian conditions.

## Observation Blank

| *Sl.No.* | *Plants to be Planted* | *Time of Planting* | *Flowering Time (Season)* | *Foliage* | *Flower* |
|---|---|---|---|---|---|
| **1.** | **Trees** | | | | |
| (i) | Flowering | | | | |
| | 1. | | | | |
| | 2. | | | | |
| | 3. | | | | |
| (ii) | Foliage trees | | | | |
| | 1. | | | | |
| | 2. | | | | |
| | 3. | | | | |
| (iii) | Avenue | | | | |
| | 1. | | | | |
| | 2. | | | | |
| | 3. | | | | |

| *Sl.No.* | *Plants to be Planted* | *Time of Planting* | *Flowering Time (Season)* | *Foliage* | *Flower* |
|---|---|---|---|---|---|
| **2.** | **Shrubs** | | | | |
| (i) | Flowering | | | | |
| (ii) | Climbers | | | | |
| (iii) | Foliage | | | | |
| (iv) | Hedge | | | | |
| (v) | Edge | | | | |
| **3.** | **Annuals** | | | | |
| (i) | Summer season annuals | | | | |
| (ii) | Winter season annuals | | | | |
| **4.** | **Biennials** | | | | |
| **5.** | **Ferns** | | | | |
| **6.** | **Cactii and Succulents** | | | | |
| (i) | Cactii | | | | |
| (ii) | Succulents | | | | |
| **7.** | **Palms and Cycades** | | | | |
| **8.** | **Water and Marshyland plants** | | | | |
| (i) | Water plants | | | | |
| (ii) | Marshyland plants | | | | |

# Exercise No. 29

## Objective

Preparation, planting and care of lawn.

## Introduction

A lawn can be defined as a green carpet for a landscape. It is a basic feature for home garden and an essential feature for any other type of garden. In a home garden, a lawn improves the appearance of the house, enhances its beauty, increases conveniences and usefulness thus adding economic value to the real estate. The lawn provides a perfect setting for a flower bed, a border, a shrubbery or a specimen tree or a shrub. Besides the material value, a lawn has its spiritiual value, too. A lawn is the source of beauty and pride and reduces tension of the mind after a day's hard work in the materialistic world. The position of a lawn is largely dependant upon the layout of a garden in relation to the house. The obvious space for a lawn should be wide open, with access to direct sunshine which may be a solitary place in a corner of the building especially in front of a rockery and a water pool.

## Materials

Spade, Khurpi, Rope, Measuring tape, Well rotten F.Y.M. or compost, Fungicides, Sprit level, Straight edge, Pegs, Grass or Seeds of grass, Roller or flat iron plate, Lawn mover.

## Methods

For making a beautiful lawn, the following steps should be taken.

## Selection and Preparation of Land

A site for lawn should receive full sunshine under shade as grass does not grow. The best situation will be the southern side and the next best is the south east and south west of the building. Most the grasses do not grow well under the drip of large trees. Therefore, it is desirable that the land piece selected of a lawn should not be closer to a big tree. Fertile loamy soil having enough humus is desirable for the growth of the lawn grasses. The soil should retain enough moisture and at the same time it should have adequate drainage. The preparation of the ground bed and drainage are essential for a lawn. A slight shape is desirable in small size plots to ensure natural drainage during rainy season. Protective drainage is essential when the plot size is large and the soil is heavy clay or water logged. Rough surface leveling by eye estimation should be done prior to digging. If during rough levelling a lot of shifting and filling of soil is necessitated, the surface soil should first be taken out and kept separately, which should be laid on the top after final levelling. After rough levelling is completed, the digging work should be done. Thorough preparation of the ground is most essential for the success of a lawn, the digging operation should be done by the trenching. At each stage of digging care should be taken that earth clods are broken and soil is pulverized thoroughly. If clods are left, the ground will take a long time to settle. During process of digging all enert material like stones, old masonry and grass roots etc. should be removed. Special care should be taken to remove the roots of nut grass (*Cyperus rotundus*).

In most parts of India, digging is done during the hot summer months (April and May). After the trenching is completed, the soil is left to dry in the scorching sun for a month. The soil should be turned up subsequently 2-3 times at weekly intervals, each time the clods of earth, if any, are broken and roots of weeds are removed. After proper digging, the soil is to be manured and levelled. In fertile soils, organic manures are not added. But poor soils are dressed with organic manures. Any freshly dugup soil takes a long time to settle. In heavy rainfall areas, the work is done by the pouring rains. In other areas, the prepared soil is watered heavily after making the bunds all along the periphery. The accumulation of water will show the area of depression which should be smoothened by shifting the soil from other places. The flooding should be repeated 2 or 3 times and between each watering the sprouted weeds should be removed. The final levelling is done with the help of levelling pegs, straight edge and sprit level. It is not always necessary to have perfectly levelled lawn. Lawn can be planted in undulated lands also and such lawns look very beautiful but there should not be any depression as then water will collect and kill the grass. Moreover, the slopes and mounds in a lawn should be gradual and artistic, simulating the nature.

## Planting of Lawn

For an excellent lawn, the grass should be of fine quality with its narrow evergreen and fast growing foliage. The most suitable grass for most parts of India is the *Doob* grass (*Cynodon dactylon*). The grass thrives well under hot and sunny weather but under shade, its growth is suppressed. The other finer varieties are *Agrostis tennis, A. canina, Festuca ovina, F. rubra,* while the cheaper mixtures consist of perennial rye grass (*Lolium perenne*) and crested dogs tail (*Cynosurus cristatus*). The wood meadow grass (*Poa nemorals*) grow well under shades of trees. If irrigation facilities exist, a lawn can be laid out any time during the year. Under Indian conditions, it is better to sow the grass seeds after one or two monsoon showers, while the grass roots are planted at the beginning of the mansoon. The following different methods of planting of a lawn are used.

(a) **Seed Sowing:** For seeding, it is important to procure a good quality seeds free from weed seeds. *Doob* grass seed is very light and fine. Therefore, proper care should be taken during sowing. Prior to sowing, the surface is scratched to a depth of 2.5 cm with the help of a garden rake. The total area should be divided into equal plots of 200 to 300 square metres to ensure even sowing of seeds. The sowing should be preferably undertaken on a windless day. The seed is divided at the rate of 500g per 200 square metres and mixed with double quantity of finely sifted soil and broad casted by hand. After sowing is completed the rake is drawn lightly twice in opposite directions to mixup the seed. It will be advisable to cover the seeds with a thin layer of finely sifted soil. The plot should be lightly irrigated at regular interval.

(b) **Dibbing:** A well matured unrooted or rooted *Doob* grass cuttings are obtained from close cut lawn on nursery or from a lawn scraping. The roots or grass thus obtained are dibbled (planted) in the ground at 7-10 cm apart when it is slightly moist. The soil is kept moist by frequent watering till the grass sprouts. Roots of *Doob* grass sprout easily and the cuttings or off shoots roots readily under moist condition and within 5-7 weeks, the grass will be ready for first cutting.

(c) **Turfing:** The quickest method of developing a lawn is by turfing. Turf is a piece of earth of about 5cm thickness with grass thickly grown over it. The pieces may be of small squares or in rolls of small widths (30 cm). The turfs must be free from weeds and consist of the required lawn grass. These should be laid closely to each other in a bounded alternate pattern, like bricks, in the already prepared ground. Any unevenness in thickness can be corrected by underpacking or removing some of the soil before putting in position. Along the joints sandy soil should be filled as packing material. The turf thus laid is made firmby a wooden beater made out of heavy block of wood and fitted with a handle. The grass is immediately watered copiously. By using this method, a lawn will be ready for use in a very short time.

(d) **Turf-plastering:** A paste is prepared by mixing garden soil, fresh cowdung and water. Bits of chopped fresh roots and stem or rhizomes of Doob grass are mixed with this paste and the paste is spread evenly on the surface of the prepared ground after moistening the soil. The paste is then covered by spreading 2 cm of dry soil and watered at regular intervals. This method is not very suitable especially in a dry and variable climate.

## Maintenance of Lawn

If the lawn is not properly maintained, it will become useless within no time. The following operations should be followed.

(a) **Weeding**: One of the main aspects of maintenance is the control of weeds. Weed is common in both new and old lawns. Therefore, as soon as a lawn is established, the weeding should be started and continued at regular intervals or whenever the weeds comeout. The frequency of weeding obviously will be more during the rainy season than in the colder months.

(b) **Liming**: If considerable growth of moss is observed, which is an indication of acidity, powdered chalk or lime should be applied at the 250 grams per square metre area. Even otherwise, application of lime once a year, except in alkaline soils, is beneficial. After liming, watering should be done.

(c) **Rolling, moving and sweeping**: The objective of rolling is to help the grass anchor itself security and also to keep the surface leveled. Rolling should be avoided when the soil is wet. Mowing is another important operation and the first thing is to obtain a good mower, which should cut the grasses evenly at a correct height. The frequency of mowing is determined by the amount of growth and will vary from season to season but the grasses should not be allowed to grow more than 5-6 cm in length during any season. Sweeping the lawn thoroughly after each mowing is essential to clean the cut grasses which might have fallen from the mower box. Sweeping is also done every morning to clean the fallen leaves and other debris material.

(d) **Irrigation**: Lawn grasses are shallow-rooted and, therefore, frequent light watering are better than copious flooding after long intervals. The frequency of irrigation varies with the season and climatic conditions of a place. Stagnation of water should not be allowed as it may kill the grasses.

(e) **Scraping and raking**: Continuous rolling, treading and mowing may result in the formation of a hard crust and the lower part of the lawn may get matted and woody. Under such situations, the grass is scraped at the ground level with the help of "Khurpi" in the months of April and May. Scraping is followed by raking to break the crust.

## Top-dressing

Fertilizing the lawn thrice in a year is adequate to maintain rich greenness and to keep the soil enriched. Application of nitrogen fertilizers, especially urea and ammonium sulphate at the rate of 0.5 and 1.0 kg/50 Sq.m., respectively, during February-March, June-July and October-November, is quite beneficial. Bonemeal at 5.0 kg/50 sq.m area will give good results. Well decomposed compost at 50 kg/50 sq.m area will be sufficient as top dressing. Spraying of micronutrients is also desirable.

## Plant Protection Measures

Some times lawn grasses are infected with certain insect-pests and diseases. Under such situation, the proper cause of the problem should be traced out and proper control measures should be adopted.

**Frost-Injury**

In North India, the grass is injured due to frost which can be avoided to a great extent if the grass is sprayed with water every evening and in the early morning after frost.

**Observation Blank**

| *Sl.No.* | *Common Name of Grass* | *Botanical Name* | *Time of Sowing/ Planting* | *Method of Sowing/ Planting* | *Time taken for First Mowing* |
|---|---|---|---|---|---|
| **A.** | **Plains** | | | | |
| 1. | | | | | |
| 2. | | | | | |
| 3. | | | | | |
| 4. | | | | | |
| 5. | | | | | |
| 6. | | | | | |
| **B.** | **Hills** | | | | |
| 1. | | | | | |
| 2. | | | | | |
| 3. | | | | | |
| 4. | | | | | |
| 5. | | | | | |
| 6. | | | | | |

**Assignments**

1. Write the common and botanical names of various lawn grasses used both under plain and hill conditions of India.
2. Select and prepare the land for planting of lawn.
3. Practice various methods of planting of a lawn.

# Exercise No. 30

## Objective

Making of herbaceous border.

## Introduction

Beds which are more in length than breadth and contain plants of heterogeneous character are known as borders. There are different kinds of borders and they are named according to the kind of plant material grown in them. Herbaceous border is a characteristics of the english garden containing hardy perennial herbaceous plants which die down to ground level after flowering but putup new growth from the roots in the next season. In actual practice, it is quite common to include bulbous plants and annuals also in the herbaceous border. The herbaceous border may be placed against a wall, fence, shrubs, a hedge or form a double border one behind another divided by a grass path. A border may be put to mark a division between one part of the garden and another.

## Materials

Spade, Khurpi, Measuring tape, Scale, Rake, Rope, Various types of herbaceous plants, Sprit level.

## Procedure

A strip of 2-3 metres wide may be selected in the background of a lawn. This strip is dug with the help of a spade. Clods are broken and the soil is prepared finely with the help of "Khurpi". Soil is raked to remove weeds. Levelling is done properly. A well rotten FYM or compost is mixed in the bed. Now the bed is ready for transplanting of seedlings. A hedge between border and lawn may be planted.

The following points should be kept in mind.

i. **Height and arrangement:** In a wide border, plants of all heights are used provided the grower can give sufficient attention on the staking of tallest plants. For a narrow width or double faced border, the maximum height should be limited to 3 m.

ii. **Grouping and colour:** Plants are arranged in bold groups. In an average sized border, each group will consist of 4-5 plants while in large borders it may vary from 5-7 plants. Each group should be planted in drifts of irregular shape. Though generally the gradation in a border is from dwarf to taller placed towards the back, it can be broaken here and there to avoid monotony or too formal a look.

The colour scheme should harmonize i.e., light shades leading the dark shades and dark shades leading to light shades. The arrangement of different colours varies with taste of individuals but basically it is an artistic job. The basic arrangement of colours in annual beds are classified as monochromatic, analogous and complementary or contrasting. The proper combination of colours is very important in annuals and same is true with plant height and time of flowering. The tall plants are planted in the background, medium types of plants in the middle and dwarf types in the front of the border. While planting, the time of flowering should be kept in mind. The late flowering types should be sown atleast a month earlier than the early flowering.

It is always desirable to draw the actual plan of planting on a paper according to a scale (1:15). The drift may be drawn showing the shape and size with the names of the plants for each drift inserted along with their height, flower colour, time and duration of flowering. The plan is then taken to the actual site and a rought idea is made as to how they will look. Then the plan is modified if the need be and actual planting is undertaken.

**Observation Blank**

| *Sl.No.* | *Name of the Species* | *Flower Colour* |
|---|---|---|
| 1. | Herbaceous winter annual borders | |
| | a. Tall | |
| | b. Medium | |
| | c. Dwarf | |

**Assignments**

1. Prepare a plan of the herbaceous border showing the colour scheme and the plant species to be planted.

# Exercise No. 31

## Objective

Making of shrubbery border.

## Introduction

Shrubbery border can be defined as a border planted with shrubs. This is an essential feature of any garden. Shrubs are of permanent nature and once planted will become a permanent feature. If properly selected, the shrubs in a shrubbery border will provide flowers throughout the year. Shrubbery borders can very effectively be used for hiding one portion of the garden from other, besides adding beauty.

## Materials

Spade, Khurpi, Measuring tape, Scale, Rake, Rope, Various types of shrub plants, Sprit level.

## Procedure

In shrubbery border, the shrubs are planted based on their height, flower colour and flowering time. In the background the tallest shrubs are plantd in one to three rows depending upon the space available, then medium height shrubs are planted. The dwarf shrubs are planted in front row. Within the rows, 4-5 plants of a species are planted according to the any of the colour schemes as has been already discussed in case of preparation of herbaceous border. The width and length of the strip will depend on the availability of the space and choice of the person. Once the every thing is planned, one should prepare the strip of shrubbery border and plant the shrubs as per procedure in case of planting of orchard trees. After planting, irrigation is required immediately in order to save the plants from wilting. The planting work may be carried out in a cloudy day or in the evening when sunlight intensity is reduced to a tolerable limit. The foliage of the shrubs should also be reduced before planting so that the plant may get least transplanting shock.

## Observation Blank

| *Sl.No.* | *Name of the Species* | *Flower Colour* |
|---|---|---|
| a. | Tall shrub species | |
| b. | Medium shrub species | |
| c. | Dwarf shrub species | |

## Assignments

1. Prepare a plan of the shrubbery border showing the colour scheme and the shrub species to be planted.

# Exercise No. 32

## Objective

Preparation of potting mixture.

## Introduction

Pots are used for various purposes in horticulture. The main uses of pots are propagation of plants, cultivation of ornamental and flowering annuals. Cultivation of plants in pots is referred as pot culture. In pot culture, operations such as potting, shifting and filling of pots are very important. Potting refers planting in pots containing soil mixture while shifting means removal of plants from pots and planting them again in the same or different pots. Plants growing for more than one year in the same pot need shifting the best time for potting and shifting is rainy season. There are many types of pots which are being utilized for growing of horticultural plants. Similarly different types of the potting mixture are used for various purposes.

## Materials

Pots, Crocks, potting mixture, Khurpi, Watering can.

## A. Selection of Pots

In India burnt porous clay pots are commonly used. These are heavy and porous and loose moisture readily. They are easily broken and their round shape is not economical for space. Continued use of these pots results into accumulation of toxic salts which requires soaking in water before reuse. Plastic pots, Fiber pots, Paper pots, metal and wooden containers are also in use. Plastic pots are becoming popular now a days in Indian nurseries. They have numerous advantages as they are nonporous, reusable, light weight and use little storage space. Small liner pots for direct rooting of cuttings, seedling propagation and tissue culture plantlet hardening and production have gained popularity.

**B. Potting Mixture**

Various potting mixtures are used as a medium. The medium must be sufficiently firm and dense and volume must be fairly constant when either wet or dry. It should be highly decomposed, easy to wet and sufficiently porous. It should be free from weed seeds, insects and various pathogens. Propagation media used in horticulture consist of a mixture of organic and inorganic components. The organic component includes peat, softwood and hardwood barks, sphagnum moss, sawdust, leaf mulch and rice hulls which oxidize and compact easily and have high C:N ratio which decreases pore-space, aeration and create nutritional problems. Therefore, a variety of mineral components like sand, grit, pumice, scoria expanded shale, perlite, vermiculite, polystyrene, clay granules and rockwool are used for improving drainage and aeration. There is no single ideal mixture. An appropriate propagation medium depends on the species, propagule type, season, propagation system, irrigation system, cost and availability of the medium components.

**C. Filling of Pots**

Select the pots of the proper size. The size will depend on the age of the plant and the purpose for which the pots are filled. Wash the pots with clean water. Place a larger crock on the drainage hole at the bottom. Put several smaller crocks to a depth of 5 to 10 cm, depending upon the size of the pot. Put 1.25-2.5 cm layer of coarse sand to check the draining of fine sand particles. Fill the pot with soil mixture. Press the mixture firmly when half filled. Continue filling upto the rim of the pot. Press the potting mixture again and finally fill and press the mixture to a point where a space of 2-3 cm is left for holding of water.

**D. Potting**

For potting of plants, digout a hole in the centre of the pot. Keep the plant in the centre of the hole with roots well distributed in all directions. Put soil all around the plant and press the soil mixture firmly. Irrigate the plant with watering can. Transfer the potted plant in a cool shady place in the nursery or in lath house.

**E. Shifting of Plants**

To avoid potbound condition, potted plants need shifting every year which is done during rainy season. In order to remove plants from the pots, the pot is held with right hand between second and third finger and the thumb along the side of the pot. The pot is then turned upside down. Tap gently on the edge of bench or on a pot to loosen the earth ball. The earth ball is then taken out from the pot carefully. Before putting the plant in a new pot, the lower finer roots and some soil mixture are removed. Irrigate the pots lightly atleast 3-4 hours before shifting in order to make operation easier.

## Assignments

1. Prepare the potting mixture for sowing of seeds and planting of cuttings.
2. Fill the pots for sowing of seeds and planting of cuttings.
3. Practice potting of ornamental foliage plants.

## Some Suggested Soil Mixtures

1. For potting rooted cuttings and young seedlings:

   Two parts sand, one part loam soil and one part leaf mould.

2. For fruit plants:

   One part sand, two parts loam soil, one part leaf mould and one part well rotted FYM

3. For sowing seeds:

   Three parts leaf mould, one part loam soil, one part sand and half part finally powdered charcoal.

4. For planting of cuttings:

   One part leaf mould, one part sand and one part loam soil.

5. For cacti and succulents:

   Loam soil two parts, leaf mould five parts, limestone gravel one part, charcoal dust one part and sand one part.

# Exercise No. 33

## Objective

Uses of flowers for different purposes.

## Introduction

Flowers symbolize purity, beauty, love, passion and tranquility, although flowers are mute beauties, they best convey the message of love, joy and affection. For example, lotus conveys purity, rose conveys love, pansy conveys thought etc. Flowers are highly esteemed for their sancity. Even at the time of birth and death of man, flowers are offered. Flowers are used for different purposes.

## (A) Religious Use

Flowers are used for different religious purposes. They are being used as offering to God. Japanese use flowers for Ikebana to express their love to God. In India, flowers are used to decorate the religious places like temple, mosques, gurudwaras and churches.

## (B) Decoration

Flowers are used for different decorations like Pandal decoration, room decoration and hair decoration. Various flower arrangements in vase are used to decorate tables, rooms (by placing at corners and elsewhere). Cut flowers like gladiolus, rose, gerbera, china aster, tuberose, etc. are used for this purposes. For decorating stages, pandal, entrance gates, loose flowers and garlands are used. Hair decoration by veni, gajra of jasmine, marigold, rose and many other flowers. The floors of the building/room are decorated with rangolies made by petals of the flowers. Wreath is made using flowers of like marigold, chrysanthemum, rose, tuberose etc.

## Industrial Use of Flowers

- ☆ **Gulkand:** Flowers like damask rose, Edward rose are used for extraction of essential oils, perfumes, attar etc. Rose is also used to make rose water and gulkand. A number of flower based industries are coming up. Many dyes are routinely prepared from various flowers like *Butea monosperma*. Flower and/or flower based products are used in confectionaries and other food industries.
- ☆ **Flower Bouquets:** On certain occasions like birthday, marriage day, welcoming/honouring degnitories, bouquets are commonly used. Bouquets are of different types and shapes. Generally, they are of either flat or round type. For making flat bouquets, white sheets are used as base. Flowers like gladiolus, tuberose etc, may be used. Many times, base is covered with ornamental leaves like Thuja, Ashok etc. The nicely arranged flowers are covered/wrapped with cellophane paper. In case of round bouquets, strong stemmed flowers with some foliage is used and wrapped with cellophen paper.
- ☆ **Button Holes:** This is another fascinating items to floral decoration. These are generally worn by males in their coat collars near the chest. Where a provision is made by crisscrossing two threads to hold the button holes. Ladies can also decorate their hair with a button hole. The most suitable flowers for a button hole are possible roses and orchids but other flowers such a carnation, camellia and daisy may also be used.

## Observation Blank

| *Sl.No.* | *Uses* | *Name of Flowers* |
|---|---|---|
| 1 | Religious | |
| 2 | Decoration<br>(i) Gajra<br>(ii) Veni<br>(iii) Pandal | |
| 3 | Industry | |
| 4 | Garland | |
| 5 | Bouquet | |

## Assignments

1. Prepare a list of flower based industries.
2. Classify the flowers on the basis of their uses.

# Exercise No. 34

## Objective

Care and maintenance of green house/poly house plants.

## Introduction

Intensive cultivation of high value, low volume horticultural crops adopting latest technologies like 'cover' or 'green house' cultivation is of late becoming crucial and vital for realising higher returns by producing high yields of better quality almost throughout the year. Green houses are framed structures covered with transparent or translucent material and large enough to grow plants under partial of fully controlled environmental conditions to get maximum productivity and quality produce.

## Materials Used

Iron frame, polythyelene plastic film, Irrigation pump, Cooling system, Exhaust fans.

## Site

It should be away from main building and trees and facilitated with proper drainage.

## Orientation

For single span orientation could be in any direction but for multi span, green house should be oriented in north-south direction to avoid continuous shading of certain portions of green house. The distinct advantages of growing crops under greenhouses are:

High productivity per unit area as the genetic potentially of the crop can be fully exploited; any crop can be grown in any season of the year depending on the demand

and market; excellent quality produce, free from any blemishes; higher extent of bud/graft take and extended period of grafting; easy to protect the crops against pest and diseases and extreme climatic conditions; supply of fresh flowers for big cities and cultivation of high valued commodities in problematic regions and extreme climate areas. The green houses are ideal for hydroponics, nutrient film techniques, for hardening of tissue cultured plants besides boasting crop productivity and for cultivation of rare ornamental species.

## Management of Greenhouse Plants

(a) **Temperature and Humidity:** In low cost naturally ventilated greenhouse, sufficient ventilation (60-70 per cent) should be provided which can be manipulated with the rollable flap. Providing sufficient irrigation will also reduce the temperature and increase the humidity to a considerable extent during summer months. Similarly during winter months by providing limited ventilation and irrigation, humidity and temperature could be regulated.

(b) **Irrigation and nutrition:** Plants can be irrigated through drip or sprinkler irrigaton systems. Fertilizers are also supplied through irrigation system. It saves the quantity of fertilizers and labour, besides improving the yield and quality of the produce.

(c) **Pests and Diseases:** Normally the incidence of pests and diseases in a green house is less as compared to the outside conditions. However, due to high crop density and congenial micro-climate in the structure, spread of the pests and diseases will be faster once there is an entry of diseased plant material or improper management of the green house. Effective prophylactic control measures to control various insect-pests and diseases are necessary to get high returns.

## Problems/Constraints of Greenhouse Cultivation

The major constraints in greenhouse cultivation are:

High initial investment required for construction and management of greenhouse. The various cultural operations in a greenhouse are intensive in nature and modified at times. Knowledge of greenhouse crop management practices is very important for successful greenhouse cultivation. Lack of organized marketing of flowers/, propagation material grown under cover is a big problem.

## Care and Maintenance

Pests are placed at intervals of 3m on both sides of the bed. Along the sides of the beds, fastened at the posts are galvanized wires or plastic string to support the plant. The side buds below the centre bud have to be carefully removed (disbudded). The raised beds have to be maintained by moving back the soils from path into the beds. Fresh soil should be added when required. Regular analysis of the soils is very important as results are used for calculating the precise amount of fertilizer requirement. Small nutrition doses but more frequently favours better growth. In carnation, before planting 4-5 layers of nets are laid out. The plants are then planted within the netting with appropriate spacing between them. The first net should be fixed at 12.0 cm above the soil. Then place the remaining nets, whose squares should be 12.5×12.5 cm or 15.0×15.0 cm over the first net. These nets are placed 15 cm apart. Proper weeding and hoeing in the plots is indispensable.

A. **Propagation of Planting material:** Green house can beneficially be used in getting a higher extent (80 to 90 per cent) of gross success and extending the period of grafting almost throughout the year irrespective of the season and ambient conditions. These greenhouses should be utilized for propagation of high value ornamental and those which are difficult to propagate under ambient conditions to fetch high and sustained returns.

B. **Growing of commercial flowers:** The low cost greenhouse could be profitably utilized for commercial cultivation of roses, Gerbera and carnations for getting higher productivity and quality flowers. A greenhouse of even 100 $m^2$ area, if utilized for crops like roses can accommodate 700-750 budded plants. An average yield of 250 flowers/$m^2$/year would give a gross return of Rs. 50,000. Thus, entire cost of construction of the greenhouse structure and cost of cultivation could be realized in one to two years.

## Observation Blank

### A. Propagation of Planting Material

| *Sl.No.* | *Species/ Variety* | *Propagating Material* | *Rootstock and Scion Used* | *No. of Grafted Plants* | *Per cent Success* | *Days taken* | *Remarks* |
|---|---|---|---|---|---|---|---|

## B. Flower Production

| *Sl.No.* | *Species and Variety* | *Per cent Success* | *Spacing* | *No. of Flowers/m²* | *Total Production* | *Time to Flowering* | *Duration of Flowering* |
|---|---|---|---|---|---|---|---|
| 1. | | | | | | | |
| 2. | | | | | | | |
| 3. | | | | | | | |
| 4. | | | | | | | |
| 5. | | | | | | | |
| 6. | | | | | | | |
| 7. | | | | | | | |
| 8. | | | | | | | |
| 9. | | | | | | | |

## Assignments

1. Draw diagrams of various types of green houses.

# Exercise No. 35

## Objective

Arranging flower show.

## Introduction

The holding of flower shows has spread widely and these are not restricted only to the capital cities of a few states but are being held in all state capitals as well as in many districts head quarters. The primary objective of flower show is to create interest among the general public to grow quality flowers and maintain beautiful gardens around their house premises. Flower shows give an opportunity to the people to know the wide range of plants than can be grown in their locality. The other advantage is that a visitor gets a chance to see all the best materials at a time in the best possible arrangement. A flower show should be a place for discussing the various garden problems and to find out ways for solving each other's difficulties. The success of a show depends much upon the co-operation between the exhibitor and the members of the flower show committee. The judging should be above criticism. Such shows are generally arranged by a society, organization, or institute devoted to floriculture. A committee meant for the purpose should prepare the schedules well in advance, prepare the rules and regulations, print the admission cards undertaken the wide publicity and invite the VIP's progressive farmers technocrats and organizers flower vases, bamboo stakes, show passes, benches and tables for exhibits should be made available to the participants. Nurserymen, seedsmen and companies, selling agricultural implements, chemicals, etc. should be allowed to open their stalls.

## Judging

The judgement is made on the variety of the specimen, the bloom, the cultivation and the impact or effectiveness of the group.

## Score Cards

There is a set pattern for judging different categories of plants and specimen blooms and marks are awarded accordingly.

**(A) Foliage plants:** Foliage plants are judged on the basis of variety of the specimen, culture, appearance or health of the plant. Points are awarded for display. The marks may be distributed as follows:

Culture: 4; Quality (including variety): 3; Appearance of foliage: 2; Display:1

**(B) Annuals:** Generally, a group of four-five plants per pot should be allowed. The marks are awarded as follows:

Quality and quantity of bloom: 4; General appearance: 3; Rarity of the specimen 2; Display and colour scheme: 1

**(C) Cut flowers:** In cut flowers, foliage as well as the bloom is seen. General qualities that are checked are uniformity, colour, size, form stem, etc. Uniformity is important as one large flower among other smaller flowers or vice-versa will affect the quality. It must be fresh and bright in colour. The stem should be strong and erect. In exhibiting a spike of flowers, the first flower should be fresh and maximum number of flowers should be open. The usual score card is:

Size and form : 5; Colour and rarity: 3; Arrangement: 2

The Rose Society of India has divided the marks for judging specimen rose blooms as follows:

Colour: 20; Form: 20; Size: 20; Substance: 20

(Texture, firmness and thickness of petals)

Foliage: 10; Stem: 10

Foliage and stem carry no marks when blooms are exhibited in boxes. But a weak neck may lose points on this count.

**(D) Floral display:** The quality of floral display is judged mainly by its aesthetic and artistic value. Hence, the superiority of a particular exhibit over the other depends largely on the personal judgement of the person rather than on specific rules and points. The marks are distributed as follows:

General set-up : 5; Colour Scheme: 3; Novel idea : 2

**(E) Bouquets:** There are mainly two types of bouquets namely posy and shower and each should be placed in separate groups while judging. The usual marks allotted on different counts are:

Size and artistic value: 5; Arrangement : 3; Novelty : 2

**Note:** In all the score cards, the marks should be multiplied by ten for the convenience of allotting marks.

## Assignments

Record the following observations in the flower show organized by department of horticulture during Farmer's Fair.

| *Sl.No.* | *Type of Competition* | *Class/Species* | *Section* | *Grading Features* |
|---|---|---|---|---|
| (A) | Foliage plant | a. | | |
| | | b. | | |
| | | c. | | |
| | | d. | | |
| (B) | Annuals | a. | | |
| | | b. | | |
| | | c. | | |
| | | d. | | |
| (C) | Cut Flower | a. | | |
| | | b. | | |
| | | c. | | |
| | | d. | | |
| (D) | Floral display | a. | | |
| | | b. | | |
| | | c. | | |
| | | d. | | |
| (E) | Bouquets | a. | | |
| | | b. | | |
| | | c. | | |
| | | d. | | |
| (F) | Button holes | a. | | |
| | | b. | | |
| | | c. | | |
| | | d. | | |

# Exercise No. 36

## Objective

Designing of nursery experiments.

## Introduction

Nursery managers and grower continually seek to improve seedling quality and cost effectiveness of their nursery practices. They face a wide variety of problems that lay themselves to research methods *e.g.*, wheather to use a new piece of equipment or a new herbicides, how dense and when to sow seed for various stock types or how to determine optimum fertilizer and irrigation regimes. Necessarily, however, recommendation to alter nursery practices are nearly always based on incomplete information which successful nursery managers evaluate in light of their experience and instincts to make sound effective decision. Statistically designed experiments often provide important information upon which nursery managers can base their decisions and calculate the degree of uncertainty associates with their conclusions.

## Designed Experiments

Given that an experiment is warranted, a designed experiment is more appropriate that an operational trial for cases benefiting from one or more of the special attributes of designed experiments (Table 1). Designed experiments are particularly useful for detailed investigation *e.g.*, establishing optimum rates of a chemical or procedure, investigating interactions among multiple factors or revealing the biological principles of a phenomenon under investigation. In these cases, the attributes of designed experiments are well worth the extra effort. Uncontrolled background variation affecting growth rate within and between nursery beds is large due to differences in fertility, soil type and irrigation. Well-designed experiments, randomized and replicated over this background variation one often required to

**Table 1: Attributes of Designed Experiment**

| *Attribute* | *Explanation* | *Comments/Example* |
|---|---|---|
| **Absence of systematic effects** | Treatment comparison are not confounded or biased due to uncontrolled (background) variations | Comparison of two fertilizer regimes would be biased by consistently applying one regime to plots in a more fertile part of nursery. |
| **Proper degree of decision** | Poor designed and large uncontrolled variation result in large experimental error and imprecise estimates of treatments effect "overdesign" results in over expenditure of effort for the necessary data. | High precision occurs when (1) experimental plots have similar background characteristics (2) experimental procedures are conducted with care and accuracy (3) a large number of replications are used and (4) the experimental design is efficient. |
| **Wide range of validity** | Inferences and conclusions will apply to the entire population of interest; replications over time and space broadens the range of validity. | Testing a herbicide for a few years in several nurseries results in broadly applicable conclusions. |
| **Quanification of degree of uncertainity** | The "reasonable shadow of doubt" accompanying experimental conclusions is quantifiable. | There is a 99 per cent chance that the new fertilizer regime results in 2+0 height increase of 3.1 to 5.2 cm above the standard regime. |
| **Point estimate** | A single number that estimates a certain quantity in the population of interest. | Treatment mean, standard deviation, minimum, maximum and range are all point estimates. |
| **Confidence interval** | For a given level of confidence, the specified range within which the quantity of interest lies. | A confidence interval on a treatment mean states with, say 95 per cent confidence that the true mean response lies between two estimated values. |
| **Hypothesis testing** | A statistical technique to accept or reject a hypothesis formulated before the experiment in light of the empirical results. | Designed experiments allow quantification of the level of uncertainity associated with acceptance or rejection. |

achieve unbiased estimates of treatment response with appropriate level of precision and range of validity.

Therefore, those designs which seem to have the most potential for use in nursery experiments will be discussed here.

(1) **Completely Randomized design (CRD)** – If the nursery manager wants to test a new weedicide at two application rates i.e., high (H) and low (L) against both a control i.e., (C = no weedicide) and the standard weedicide (S = standard), that each of the four treatments is replicated 5 times in 20 plots. In this case treatments are assigned totally at random. Each nursery bed is not assigned one complete replicate of the experiment.

(2) **Randomized complete block design (RCBD)** – Each nursery bed is assigned one complete replicate of the experiment and randomization is restricted to the allocation of treatments within a block. If nursery beds are scattered representing a wide variety of conditions, plots within a bed should be more similar to one another than to those from different beds.

(3) **Split Plot design -** With factorial treatments, it sometimes happen that one of the factors calls for larger plots than the other. Split plot designs are at their most useful with provocative factors on the main plots and the substantive one on the sub plots. For example, the minimum size of plots for irrigation treatments and sowing dates are necessarily larger than that for different sowing densities and seed source. In this design, the treatments in main plots are compared with less precision than those in the sub plots. A split plot design for two factors calls for assigning treatments of one factor to larger plots known as main plots and then splitting each main plot into enough sub plots to accommodate one replicate of each treatment level of the second factor. Split plot designs can be extended to multiple factors at both the main plot and sub plot level and even to splitting the sub plots into split-split plots.

(4) **Criss-cross (or Strip plot) designs** – These designs are related to the split Plot design. These designs are really row and column designs. These designs are better suited for two factor experiments when both factors require large plot sizes and when the interaction between the two factors is desired to be measured more precisely than the effects of either of two factors. If the source of error are effectively associated with the rows and columns and there is no interaction between them, the interaction will be better estimated than either of the main effects and that is often desirable. An important advantage lies in factors being applied to log narrow areas that go through from headland to headland. With some treatments, for example, depth of ploughing and line irrigation, that can be invaluable. It can help also with the harvesting of different varieties. On the other hand, although the design is in rows and columns, it no longer absorbs its own headland effects.

(5) **Nelder design** – In some circumstances, it is proper to use non-random, systemic designs. Where one of the treatment factors is a quantitative variable, such as, planting density and neighboring effects are likely to exits, systemic

designs can remove the need for internal guard rows. Systemic designs have been in use in intercropping and horticultural crops for quantifying the yield-plant population/spacing relationships. Nelder design for plant density can be adapted to deal with a system involving more than one species in a number of ways. Where it is desired to keep the ratio of the two species constant but vary the overall density a certain proportion of the 'spokes' (radii) can be planted with each species. Planting alternate spokes in a fan design with different crops, for example, will produce the simple design. Where the density of only one of the species is to be varied it is planted in the positions indicated by the fan design while the other species is planted on the same land at its usual spacing.

(6) **Neighbour – balanced designs** – In nursery experiments, randomized designs often require extensive border rows or plots, particularly if neighboring plots have very different treatments and these are seen as wasted space and efforts. An alternative approach is to restrict the randomization of a randomized design to give it some of the advantage of a systemic designs, but not to the extent that the analysis of variance can no longer be applied. This approach is applied to 5×5 Latin square (LSD). In this nearest neighbour designs; each treatment occurs next to each other treatment on four occasions (i.e. to the right, left, above and below) on four occasions. Thus any deleterious or beneficial effect of one treatment on its neighbour should balance out. This may permit some reduction in the area needed for guard rows, Analysis of these designs can be the usual method for Latin square or special nearest-neighbour methods can be used. Unfortunately, neighbour-balanced designs do suffer from two serious practical disadvantages. Firstly, in order to use them, the exact nature of the neighbour effects must be known at the design stage of experiment. Secondly, neighbour-balanced designs can be constructed only for certain situations and even when they can be constructed a valid randomization scheme is not always possible.

(7) **Beehive designs** – In most experiments, plants are arranged in rows but, if the interactions between individual plants are of interest, a hexagonal grid can be useful and such arrangements can be made in competition experiments. These designs are termed as beehive designs. In two different species, different plants have different number of neighbours of the other species, thus allowing the measurements of competitive effects. Hexagonal designs can be easily adopted to the situation where more than two species are involved and provide a greater number of inter-specific interfaces than in arrow layout. Planting on triangle ensures that each non-boundary or internal hill has equidistant nearest neighbour which together with the honeycombing property of the implied hexagonal grid leads to layouts requiring less material and having more scope than designs for the same purpose on a squire lattice.

## Choosing an Experimental Design

Applying the concepts of randomization, replication and error (local) control to nursery field experiments produce frequent use of only a few common designs. CRD's in which the random assignment of treatments to plot is unrestricted are not common in nursery experimentation. A field grouping of experimental plots almost always exist such that plots within a group are more similar to each other with respect to water drainage. Proximity to irrigation and or soil characteristics than are plots from different groups. In these case, RBD's are the most commonly used in nurseries. Moreover, this design is relatively easy to layout and analyze and is relatively insensitive to the accidental loss or destruction of a plot or two. Latin Square designs are used, though less frequently, when bidirectional field gradients exist. For multifactor experiments, especially those investigating irrigation, sowing date or wrenching, Split Plot design are common. Criss cross (or Strip plot) designs are better suited for two factors experiments when both factors require large plot size and when the interaction between the two factors is desired to be measured more precisely than the effects of either of two factors. Beehive designs are used where interactions between individual plants are of interest. Hexagonal grid leads to layouts requiring less material and having more scope than designs for the same purpose on a squire lattice. Nelder design can be used to study the effect on yield of a large range of plant densities, each growing at a large range of pattern of plant arrangement. Conventional RCBD will require larger area in which half of the plants would be guards. Nelder designs require small area. Further, guard plants would only be needed around the outer edge of a group of plants arranged in a systemic manner.

## Assignments

1. Calculation of dimensions for a Nelder design.

   **Given :** Number of densities (N) = 8

   Area per plant = $A_1$ to $A_N$ = $1m^2$ to $10m^2$

   Rectangularity $\tau = 1$ (square arrangement)

   $\Pi$ radians = $180^0$

   X = $30^0$ 12′ = half angle of the fan

2. **Calculate the**

   (i) $\alpha$ - a constant governing the rate of change of spacing.

   (ii) $\theta$ - angle between the radii.

   (iii) $r_0$- distance of the first plant from the centre.

   (iv) $r_1, r_2, r_3$ ———————— $r_{N+1}$ from the relation $r_{n+1} = \alpha\, r_n$.

   (v) L = length of the rectangular plot.

   (vi) B = breadth of the plot.

## Formulae

1. $(2N\text{-}2) \log \alpha = \log A_N - \log A_1$.
2. $\theta = \tau\,(\acute{a}\text{-}1)/a^{1/2}$.
3. $r_0 = [2A_1/\theta\,(\alpha^3 - \alpha)]^{1/2}$.
4. $L = 2r_{N+1} \sin X$.
5. $B = r_{N+1} - a$.
6. $A = \theta \times r_0$.

# Exercise No. 37

## Objective

Media preparation, testing and potting of fruit plants.

## Introduction

Soil is composed of materials in the solid, liquid and gaseous states. For optimum plant growth, all these materials must exist in the proper proportions. The solid portion of a soil is comprised of both inorganic and organic forms. The inorganic part of soil consists of the residue from parent rock after decomposition resulting from the chemical and physical processes of weathering. The organic portion of soil consists of both living and dead organisms.

(i) Insects, worms, fungi, bacteria and plant roots generally constitute the living organic matter.

(ii) Animal and plant life in various stages of decay make up the dead organic materials.

The liquid part of soil, the soil solution, is made up of water containing dissolved minerals in various quantities as well as dissolved $O_2$ and $Co_2$. The gaseous part of the soil is important to good plant growth. In poorly drained, water logged soils, water replaces the air, thus depriving plant roots, as well as certain desirable aerobic microorganisms, of the $O_2$ necessary for their existence. The texture of a soil depends upon the following components.

(i) Sand : 0.05 to 2 mm particle diameter.

(ii) Silt : 005 to 0.002 mm particle diameter.

(iii) Clay : ess than 0.002 mm particle diameter.

The principal texture classes are:

(i) Sand.

(ii) Sandy loam –(70 per cent sand + 20 per cent silt + 10 per cent clay).

(iii) Loam.

(iv) Silt loam.

(v) Clay loam (35 per cent sand +35 per cent silt + 30 per cent clay).

## For Good Result the Following Characteristics of the Medium are Required

(i) It must bed sufficiently firm and dense to hold the plants. Its volume must be fairly constant when either wet or dry that is excessive shrinkage after drying is undesirable.

(ii) It must retain enough moisture so that watering does not have to be too frequent.

(iii) It must be sufficiently porous so that excess water drain away permitting adequate penetration of $O_2$ to the roots.

(iv) Media must be free from weed seeds, nematodes and various pathogens.

(v) It must not have a high salinity level.

(vi) Media should be capable of being pasteurized with steam or chemicals without harmful effects.

(vii) It should provide adequate nutrients in situations where plants are to remain for a long period although supplementary slow release fertilizers are frequently recommended.

## Other Component of Soils

A. **Sand:** It is used mostly in combination with organic materials. Sand is the heaviest of all media used, a cubic foot of dry sand weighing about 45 kg.

B. **Peat:** This consists of the remains of aquatic, marsh, bog or swamp vegetation which has been preserved under water in a partially decomposed state. There are three types of peat.

   (i) *Moss peat*: This is the least decomposed of the 3 types and is derived from sphagnum or other mosses. It varies in colour from light tan to dark brown. It has a high moisture holding capacity (15 times its dry weight). It contains one per cent nitrogen, no phosphorus or potassium.

   (ii) *Reed sedge peat*: This consists of the remains of grasses, reeds, sedges and other swamp plants. Colour of this peat is reddish brown to almost black. The pH ranges from about 4.0 to 7.5 and its water holding capacity is about 10 times its own dry weight. This peat is not used for horticultural purposes.

   (iii) *Peat humas*: This is in such an advanced state of decomposition that the original plant remains can not be identified. It is dark brown to black in colour with a low moisture holding capacity but with 2.0 to 3.5 per cent nitrogen.

C. Sphagnum moss: It contains such small amount of minerals that plants grown in it for any length of time with added nutrients. Sphagnum moss has a pH of about 3.5 to 4.0.

D. Vermiculite.

E. Perlite.

F. Pumice.

G. Rock wool.

H. Shredded bark and wood shavings.

I. Synthetic Plastic aggregates.

J. Compost: It is defined as the biological decomposition of bulk organic wastes under controlled conditions, which take place in piles or bins. The process occurs in 3 steps.

   (i) An initial stage lasting a few days in which decomposition of easily degradable soluble material occurs.

   (ii) Lasting several months, during which high temperature occurs and cellulose compounds are broken down.

   (iii) Final stabilization stage when decomposition decreases, temperature lower and microorganisms re-colonize the materials.

## Mixture for Container Growing

Container growing of young seedlings and rooted cuttings has become an important alternative for field growers. For this purpose special growing mixtures are needed. To provide uniform potting mixtures of better textures, sand and some organic matter, such as peat moss or shredded bark, may be used alone or added to a loam soil. In preparing these mixtures, the soil should be screened to make it uniform and to eliminate large particles. Preparation of the mixture should preferably take place at least one day prior to use. The mixture should be just slightly moist at the time of use so that it does not crumble; on the other hand, it should not be sufficiently wet to form a ball when squeezed in the hand.

## Potting Mixtures Containing Soil and Mixed by Volume are:

**A. Heavy soils such as clay loam or clay:**

Soil – one part.

Perlite or sand – two part.

Leaf mold or Peat moss – two part.

**B. Medium soil such as silt loam**

Soil – one part.

Perlite or sand – one part.

Peat moss or leaf mold – one part.

**C. Light soil such as sandy loam**

Soil – one part.

Peat moss – one part.

For each (35 litres) of forgoing mixes add:

224 g dolomitic lime stone.

280 g 20 per cent superphosphate.

For example, one successful mix for small seedlings, rooted cuttings, and bedding plants consists of one part each of shredded fir or pine bark, peat moss, perlite and sand. To this mixture is added gypsum, superphosphate, dolomite lime and potash. Nitrogen and potassium are added subsequently in the irrigation water or as a top dressing of slow release fertilizer. Many commercial ready-mixed preparation are available in bulk or bags and are widely used by nurseries and home gardeners. Some mixes, already filled into cell packs, seed treys, or pots are available, ready to planted.

**Assignments**

1. Give potting mixtures of heavy soils, medium soils and light soils.
2. Give the mixtures for container growing.

# Exercise No. 38

## Objective

Water requirement of nursery plants and irrigation techniques.

## Introduction

Water is one of the most important inputs essential for the production of crops. Plants need it continuously during their life and in huge quantities. It profoundly influences photosynthesis, respiration, absorption, translocation and utilization of mineral nutrients and cell division besides some other processes. Both its shortage and excess affect the growth and development of a tiny seedling directly and consequently, its yield and quality. As a first step in the proper design of an irrigation scheme it is necessary to know the crop water requirements. The data required can be obtained by measuring the amount of water used by different crops under field conditions. Direct measurement procedures are laborious and time consuming. Consequently a large number of estimation methods have been developed. The five most widely known and used are the Blaney-Cridle, Radiation, Penman, Penman-Monteith and Pan Evaporation methods. In general terms, the crop water requirement is equivalent to the rate of evapotranspiration necessary to sustain optimum plant growth. More specifically, the water requirement is defined here in as "the rate of evapotranspiration of a disease free crop growing in a field of not less than one hectare under optimal soil conditions assume adequate fertility and water to achieve the full production potential of the crop under the prevailing environment". The crop water requirement is designated as ET crop and is expressed in millimeters per day (mm/day) or litre per day per plant (1pd/plant).

The scientific utilization of water resources for nursery production involves the consideration of the suitability of water for irrigation and then the planning of water management practices and water application techniques. Irrigation water application techniques may be broadly classified into two following groups:

1. Conventional methods, and
2. Advanced methods.

## Conventional Method

Conventional methods of water application have been practiced in India since time immemorial. It is also known as surface irrigation or gravity irrigation in which water applied is distributed by means of open surface flow. This may be further divided into following groups:

(1) Flood irrigation
(2) Check basin irrigation
(3) Ring basin irrigation, and
(4) Furrow irrigation

These methods are not suitable for nursery growing due to their several ill effects. Water in all these methods are applied freely and after a considerable interval of time, thus, causing flood and moisture stress alternatively to a tender plant. Therefore, these methods are not suitable for nursery raising.

## Advanced Methods

To avoid flooding and severe moisture stress on plant, modern methods of irrigation have been developed. It may also be further divided into three categories:

a. Sprinkler irrigation.
b. Micro irrigation such as (i) drip irrigation, (ii) microsprinkler irrigation, (iii) microsprayer irrigation, (iv) bubller irrigation and (v) pop-up irrigation.
c. Capillary irrigation.

## Assignments

1. Calculate the water requirement of orange, mandarin, lemon and guava of ten month old plant in the nursery?
2. Give the water requirement of hibiscus, rose and bougainbillia plants in the nursery. Give the suitable methods of irrigation in the nursery.

# Exercise No. 39

## Objective

Fertilization of nursery stock.

## Introduction

To avoid stress and poor development during propagation, it is important that the stock plants be maintained under optimal nutrition-prior to harvesting propagules. During propagation, nutrients are generally applied to stock plants are pre-incorporation into the propagation medium or with soluble fertilizer applied after roots are initiated. However, scantly literature is available, many on the three primary elements *Viz.*, N, P and K and in few cases with the trace elements.

## Manuring Schedules for Mango, Lemon, Guava and Litchi at Nursery Stage in Subtropical Fruits as Suggested

| *Crop* | *Recommendations* | *References* |
|---|---|---|
| Mango | 1. Liquid manure @ 225 g/plant at fortnightly intervals starting 3 months after planting<br>2. Ammonium sulphate or calcium ammonium nitrate at the rate of 22 to 27 kg N/ha once or twice. | Cheema *et al.*, (1954) |
| Guava | 1. 25 g N and 50 g P/plant | Tiwari and Tiwari (1993) |
| Lemon | 1. Combined application of 100-150 kg N and 75-100 kg K/ha | Mishra (1990) |
| Litchi | 1. FYM 10 kg + 0.3 kg CAN + 0.2-0.6 kg super phosphate + 0.05-0.15 kg MOP/plant | Nijjar (1972) |

## Method of Application

Nutrient element can enter plants through the roots, the foliage and in some cases through the bark.

## Surface Application

Band placement along the rows is most commonly used by nurserymen. It is an effective and efficient method for the application of nitrogen.

## Foliage Application

Fertilizers can be applied to the leaves of plants since leaves can effectively absorb applied nutrients. The nutrient ions apparently penetrate the cuticle either directly or through thin permeable areas associated with guard cells and trachoma. The process is most active when the leaves are warm and the leaf surfaces remain wet for fairly long periods. Nitrogen is absorbed very rapidly, within a few hours, when applied to the foliage. K, Ca, Mg, Mn and Zn are absorbed fairly rapidly in less than 24 hours whereas it can take days for S, Fe and Mo to be absorbed. Ca, Mg, Mn and Zn deficiency can be readily corrected with one or two sprays during the growing season.

## Fertigation

Fertigation implies the application of fertilizer or chemicals through the irrigation system.

## Assignments

1. Calculate the requirement of N, P and K for mango, guava, lemon and litchi in ten month old plants in the nursery?
2. Calculate the zinc and boron of one year old plants of mango and lemon in the nursery through foliar application?

# Exercise No. 40

## Objective

Testing of nursery soil and plant tissue of fruits.

## Introduction

### (1) Testing of Nursery Soil

Addition of relatively large amounts of organic materials and mineral fertilizers, along with removal of nutrients in the crop and by leaching and other losses, result in considerable fluctuation in nutrient levels in nursery soils. Because nursery soil require different types of management than orchards soil, therefore, much thought be given to soil and site factors that make for efficient operation in the production of quality seedlings. A soil test provides information on the fertility status of soils that is particularly useful prior to seedbed preparation. Soil samples should be collected from seedbeds 4 to 6 weeks prior to seedling to allow adequate time for sample analysis and application of necessary fertilizers. A composite soil sample should be taken from each nursery management unit that differ in soil type, past management, or productivity. The composite sample should consist 12 to 15 sub samples (Cores) taken at random over a uniform area. Cores are normally taken to a 15-20 cm depth with a 2.5 cm diameter, soil to be collected in a clean container. After air drying and mixing, about 250 g is packaged, lebelled and submitted to a laboratory for testing. Chronological records of soil test results fertilizer treatments, fumigation, seedling date and rate, irrigation and other cropping practices are useful for diagnosing problem areas and planning for future. Soil fertility tests should be carried out annually. Following soil tests are usually done in the laboratory for nursery soil.

1. Soil Texture.
2. Soil reaction/pH.
3. Electrical conductivity/Soluble salt.
4. Soil organic carbon.
5. Available N, P, K and S.
6. Available micronutrients, Zn, Fe, Cu, Mn, B and Mo.

1. **Soil Texture:** Loamy sands to sandy loam are generally preferred texture for nursery soil. Coarse textured sands are acceptable for nursery but coarse texture place special demands on management of water, fertility and organic matter. Soils containing 30-40 per cent silt plus clay are to be avoided. It determined by Bouyoucos Hydrometer method. Soil depth and texture are prime importance for nurseries. A soil depth of 1 to 1.5 meter is normally desired.
2. **Soil pH:** Soil pH indicates whether soil is acidic, neutral or alkaline in reaction. Since crop growth suffer too acidic or alkaline range, therefore, near neutral pH, i.e. 6.5-7.5 is optimum for most of crop plants. However, optimum pH for nursery soil is between pH 5.2 to 6.2. In the laboratory, pH is measured by glass electrode pH meter. Now a days digital pH meters are available for measuring pH in the field.
3. **Electrical Conductivity:** It is more important that nursery soil not contain high concentration of soluble salts, free carbonates or toxic materials. Measure of electrical conductivity will provide the concentration of soluble salts in the soil at any particular temperature and it is measured by E.C. meter in the laboratory.
4. **Soil Organic Carbon:** The maintenance of reasonable level of organic matters is particularly important in sandy nursery soil to retain fertilize elements against leaching and to buffer the soil against rapid change in pH. In the laboratory soil organic carbon is measured by Walkley and Black method (1934).
5. **Available Nutrients:** Available major nutrients in soil are determined by the procedures as described by Jackson (1973). Kalrad and Maynard (1991) and Page, Miller and Keeney (1982). While available micronutrients such as Zn, Fe, Cu and Mn are determined by procedures as described by Lindsay and Narvell (1978). Available boron is determined by Hot water method as described by Berger and Troug (1939) and available Mo by oxalate method. Available sulphur in soil is determined by Williams and Steinbergs (1959) method.

### (2) Testing of Plant Tissue

Two genral types of plant analysis have been used. One is tissue test on fresh tissue in the nursery and other is tissue analyses performed in laboratory. Fresh tissue tests are important in diagnosing the nutrient needs of nursery plants in fields. The concentration of nutrients in the cell sap is usually a good indication of how well the plant is supplied at time of testing. These semiquantitative tests are intended mainly for verifying or predicting of N.P.K.S. and several micronutrients. In general, the conductive tissue of the latest mature leaf is used for testing, while immature leaves at the top of plant are avoided. The most critical growth stage for tissue testing is at bloom or from bloom to early fruiting stage.

## General Procedure

In this, the plant part is chopped and extracted with reagents. For extraction of the easily soluble nutrients two solutions are used.

1. Morgan reagent: Glacial acetic acid preferred by means of sodium acetate to pH 4.8. This is used for extraction of NO3, PO4, Ca, Mg, K and C1, and of Mn and Al for toxicity.
2. Hydrochloric acid: It is used for the extraction of Fe and Zn when present in toxic concentration. A few points related to tissue tests are as follows:
   (a) It is ideal to follow the uptake of nutrients through the season by testing five or six time.
   (b) There can be two peak periods of nutrient demand i.e., maximum vegetative growth and second is during reproductive stage.
   (c) Comparison of plants in field is helpful test plant from deficient areas and compare them with plants of normal area.
   (d) Plants vary, thus test 10 to 15 plants and average the results.

## Total Analysis

Total analysis is performed on the whole plant or on plant part. Leaf is the principal site of crop physiological activity, showing, symptoms of nutrient stress and hence is an ideal index tissue. Recently mature-1 leaf is the most suitable index tissue which represent nutrient status of plants. For subtropical fruits following table should be consulted.

| *Crops* | *Plant Part, Age and Position* | *Sample Size* | *Optimum Leaf Nutrient Norms* | | |
|---|---|---|---|---|---|
| | | | *N%* | *P%* | *K%* |
| Mango | Collect 4-5 month old leaves from middle part of shoot | 30 | 0.84-1.53 | 0.064-0.147 | 0.52-1.10 |
| Guava | Third pair of leaf from apex. August/December | 30 | 1.60-2.40 | 0.15-0.30 | 1.30-1.70 |
| Grape | Petiole from fifth leaf | 200 | 0.87-1.60 | 0.30-0.65 | 2.00-3.00 |
| Lemon | Basal 4 month old leaf from current season growth | 50 | 2.20-2.70 | 0.15-0.30 | 1.00-2.00 |

| *Crops* | *Plant Part, Age and Position* | *Sample Size* | *Optimum Leaf Nutrient Norms* | | |
|---|---|---|---|---|---|
| | | | *N%* | *P%* | *K%* |
| Ber | Fifth leaf from tip of secondary and testiary shoot in June | 50 | 1.50-2.20 | 0.14-0.45 | 1.60-2.00 |
| Litchi | Second pair of leaflets from tip from autumn flush, 6 month old | 40 | 1.50-1.75 | 0.100-0.200 | 0.40-0.70 |

**Processing of Plant Samples**

1. **Washing:** After sample collection, the fresh tissues should decontaminated from dust and other foreign material by adopting following procedure (a) 0.2 per cent liquid detergent (b) N/10 HCL solution (c) Deionized water. The fresh tissue should be washed in sequence in detergent solution, dilute HC1 and deionized water. The extra moisture is wiped out, the sample is placed in new paper bag.
2. **Drying:** After washing, all samples should be dried in hot air oven at temperature of 65° ± 5°C.
3. **Grinding:** The over dried samples should be ground in a grinder filtered with sieve of 40 mesh. After grinding plant material should be mixed thoroughly and transfer to tissue paper bags, labeled clearly and stored in room meant for this purpose.

**(3) Analysis of Plant Samples**

1. **N determination in plant:** Total N in plant sample is determined by micro Kjedhal method as described by Jackson (1973). Whole procedure is completed in three steps. 1. Digestion of sample in sulphuric acid at a temperature 360⁰-410⁰C, the samples are cooled and distilled with 40 per cent NaOH. The distilled $NH_3$ is quantitively absorbed in boric acid and titrated against standard acid.
2. **Plant analysis for P, K and micronutrients:** For release of mineral elements from plant tissues, dry ashing and wet oxidation are two procedures. Dry ashing is carried out at an ignition temperature of 550 to 600⁰C followed by its extraction in dilute HC1. Wet oxidation employs oxidizing acids like $HNO_3$- $H_2SO_4$- $HC1O_4$ triacid in ratio of 10:1:4 or diacid in ratio of 10:4. Use of $HC1O_4$ avoids the volatilization loss of K and provides a clear solution while $H_2SO_4$ helps completion of oxidation with $HNO_3$. Direct contact of $HC1O_4$ with plant samples might lead to explosion and fire and hence predigestion of samples in $HNO_3$ is preferred.

After digestion of plant material definite volume is made by diluting the digest with distilled water. In this material (aliquot) total P is determined by spectrophotometer by measuring yellow colour intensity as described by Jackson

(1973). Total potassium is determined by flame photometer and micronutrients such as Cu, Zn, Fe and Mn by direct automization of aliquot in atomic absorption spectrophotometer.

**Assignments**

1. Collect the soil samples for one year old plants in the nursery?
2. Collect the leaf samples of mango, guava and litchi for one year old plants in the nursery for nutrient analysis?

# Exercise No. 41

### Objective

Use of plant growth regulators in nursery production.

### Introduction

Plant hormones are organic compounds other than nutrients, produced by plants which in low concentrations, regulate plant physiological processes. They usually move within the plant from site of production to a site of action. On the other hand plant growth regulators/plant bioregulators/plant biomodifiers are either plant hormones or synthetic compounds that modify plant physiological processes. They regulate growth by mimicking hormones, by influencing hormone synthesis, destruction, or translocation, or (possibly) by modifying hormonal action sites. In distinguishing between plant hormones and plant growth regulators, it may be said that all hormones regulate growth but not all growth regulators influence various processes of plants. Growth regulators can be used for following purposes:

1. Propagation.
2. Inhibition of abscission.
3. Prevention of bud dormancy.
4. Growth control.
5. Control of germination.
6. Breaking of dormancy.
7. Seed germination.
8. Seedling growth.

The use of various plant growth regulators in nursery production are as follows:

## A. Seed Propagation

### 1. Prevention of Germination

Abscissic acide has been described as an important compound in controlling embryo development and preventing germination. Therefore, it may be used for enhancing storage life of some viviparous plants.

### 2. Breaking of Dormancy

The presence of cytokinins and gibberellins are associated with embryo enlargement phases of development, whereas inhibitors (particularly ABA) along with dehydration have been associated with the maintenance of the embryonic phase and prevention of premature germination. Germination initiation, breakdown of storage products and seedling growth have been associated with gibberellins, whereas cytokinins are effective in neutralizing the inhibitors. While there are many kinds of natural gibberellins in plants, $GA_3$ is one of the widely used for exogenous application. $GA_3$ treatment can overcome physiological dormancy in various seeds and stimulate germination in seeds with dormant embryos. In seed testing laboratories, the germination medium is usually moistened with 500 mg/1 $GA_3$, although range of 200-1000 mg/1 are used. At concentration above 80 mg/1, addition of a buffer solution is recommended. Salt formulation of GA is readily water soluble. For large seeds, a 12 h soak in 500-1000 mg/1 is recommended. Cytokinins stimulate germination of some kinds of seeds by overcoming germination inhibitors. A commercial preparation, kinetin (6-furfurylaminopurine), is available. Dissolve first in a small amount of HC1, then dilute with water. Other available synthetic cytokinins are BAP (6-benzylamino purine) and PBA (6-benzyl-amino)-9-(2-tetrahydropyrany1)-9H-purine. These are more active for higher plants than is kinetin. These materials stimulate germination and overcome high-temperature dormancy of certain kinds of seeds, where seeds were soaked in 100mg/1 kinetin solution for three hours. For larger seeds, BAP at 200 mg/1 for 24 hours has been used. Cytokinins are sometimes effective in promoting germination when used in combination with gibberellic acid and ethylene producing compounds.

## B. Vegetative Propagation

### 1. Cutting

Various classes of growth regulators such as auxin, cytokinin, gibberellins and ethylene as well as inhibitors such as abscissic acid and phenolics influence root initiation. Of these, auxin have the greatest effect on root formation in cutting.

### Auxin

Indole-3-acetic acid (IAA) was identified as a naturally occurring compound having considerable auxin acitivity. Two synthetic materials, IBA and NAA were even more effective than naturally occurring IAA for rooting. It has been confirmed that auxin is required for initiation for adventitious roots on stems and indeed, it has been shown that division of first root intial cells are dependent upon either applied or endogenous auxin.

### Cytokinin

Generally a high auxin: low cytokinin favours adventitious root formation and low auxin: high cytokinin favours adventitious bud formation.

### Methods of Application of Plant Growth Regulators

There are three methods of use of plant growth regulators in cuttings:

(a) Commercial powder preparation.
(b) Dilute solution soaking methods.
(c) Concentrated solution dip methods (quick-dip method).
(d) Preparation of IBA or NAA talc formulation.

### (a) Commercial Powder Preparation

Complete direction come with the commercial materials, together with a list of plants that are likely to respond to the particular preparation. Woody, difficult-to-root species should be treated with higher concentration of preparations, whereas, tender, succulent of and easily rooted species should be treated with lower strength materials. Fresh cuts should be made at the base of the cuttings shortly before they are dipped into the powder. The operation is faster if a bundle of cuttings is dipped at once rather than each cuttings individually, being sure that the inner cuttings in the bundle receive as much powder as those on the outside.

### (b) Dilute Solution Soaking Methods

The basal part 2.5 cm of cuttings is soaked in a dilute solution of the material for about 24 hours just before they are inserted into the rooting medium. The concentrations used vary from about 20 ppm for easy to root species to about 200 ppm for more difficult to root species. During the soaking period, the cuttings should be held at about 20°C, but not placed in the sun. This is a slow, cumbersome technique and is not commercially popular.

### (c) Concentrated Solution Dip Methods (Quick-dip Method)

In quick dip method, a concentrated solution varying from 500 to 10,000 ppm or higher of the root-promoting chemicals, in 50 per cent alcohol is prepared, and the basal ½ to 1 cm of the cutting are dipped in it for short time (usually three to five seconds), then the cuttings are inserted into the rooting medium. Cuttings are most efficiently dipped as bundle, not one by one. Many propagators prefer to use this method for commercial propagation.

### Preparation of IBA and NAA Quick-Dip in 50 per cent Alcohol

Concentration : 500 to 10,000 ppm.

Duration of basal dip : 3 to 5 second.

| Final Concentration (ppm) | Auxin (per litre of solution) | |
|---|---|---|
| | Mg | Grams |
| 500 | 500 | 0.5 |
| 1000 | 1000 | 1.0 |
| 5000 | 5000 | 5.0 |
| 10000 | 10000 | 10.0 |

1. To make a 10,000 ppm (10,000 mg/1 or 1 per cent solution of auxin, dissolve 10 g of auxin in 15 to 20 ml of alcohol (ethyl, isopropyl or methyl) then top to 1000 ml (1 litre) with 50 per cent alcohol.
2. To make 1 litre (1000 ml) of a 1000 ppm auxin solution from the 1 per cent stock solution, add 100 ml of the 1 per cent stock solution and 900 ml of 50 per cent alcohol.
3. With IBA or NAA follow the same procedures except that water is used as the solvent.
4. Avoid precipitation problems with auxin solutions by using distilled or deionized water, not tap water.
5. Label solutions and colour code, different concentrations with food dyes, which can be purchased from supermarkets.

## (d) Preparation of IBA or NAA Talc Formulation

Follow the same concentration procedures as with liquid preparation (substituting 1g of talc for 1 ml of final liquid). To make 200 g of a 5000 ppm auxin talc, dissolve 1g of auxin in 50-100 ml of alcohol or acetone and then add 200g of talc (baby powder or talc from the local formacy will suffice). Thoroughly stir the slurry of talc-auxin-alcohol and allow to air dry in a well-ventilated area. Rather than making serial dilutions from the highest stock solution as is done with concentrated solution dips, it is more accurate and easier with talc formulations to make up each concentration individually.

## Propagation by Means of Layering

Commercial auxin preparation are used for root initiation in layering process.

## Preparation of Lanolin Paste

Follow the same procedures as with liquid preparation (substituting 1g of lanonin for 1 ml of final preparation). To take 200 g lanolin in a beaker and melt the lanolin by placing the beaker on a hot plate. When the lanolin is totally melted, remove the beaker from hot plate. Make 200g of a 500 ppm auxin-lanolin paste, dissolve 1 g of auxin in 50 ml of alcohol or acetone and then add in melted lanolin. Thoroughly stir the lanolin-auxin-alcohol mixture and allow to cool down and then keep it for 24 hours in refrigerator to evaporate alcohol from the paste.

## Assignments

1. Prepare a one litre solution of 500 ppm of IBA?
2. Prepare a one litre solution of 100 ppm of NAA?

# Exercise No. 42

## Objective

Protected nursery production in fruits.

## Introduction

Though India enjoys wide range of climatic conditions and various agro-climatic zones favours the production of various kinds of seeds and other prapagules of subtropical fruit crops in one or in other zone, however, there is a wide seasonal, monthly, weekly or even daily fluctuation in temperature, light intensity and humidity which not only affect the quality of propagules but also limits their availability for particular period of the year. This necessitates the modification of the environment to the congenial limits by artificial management of environmental conditions. Due to initial high cost of construction, the protected nursery production has not so far been adopted by many growers. However, a few private nurserymen have put up some structures for rapid multiplication of ornamental crops. With the increased awareness about the protectd cultivation and green houses, number of nurserymen and seed growers are coming forward for setting up these facilities for raising ornamental plants.

## Climatic Zones of India in Relation to Protected Structure

India can be divided into three main zones for plasticulture purposes:

1. Wet humid as of south and coastal areas characterized by frequent tropical rains and warm temperatures.
2. The central plains and desert-areas characterized by moderate rain and wide fluctuating temperature in winter (0°C) and summer (40°C).
3. Hill areas of north and north-eastern areas having moderate to high rainfall with freezing temperature in winter to moderate temperature in summer.

Therefore, for different regions, different types of structures and covering materials will be required. For warm humid areas, low plastic tunnels and plastic houses will be highly useful for raising nurseries, transplants and bedding plants. Hence ground to ground Quonset type polyethylene structures of 5-20 meters are highly suitable with side folding. Under central plains and desert areas, structures like polyhouses having fan pad cooling system will be highly useful. Due to prevalent extremely low temperature in winter, greenhouses made of polycarbonate of single layer, having heating system and $CO_2$ enrichment will be useful in raising nursery and bedding plants. There are three common types of greenhouses, *viz.* Gable, Quonset and ground to ground. Glass, plastic, polyethylene, polyvinyl fluoride (Tedlar), polyvinyl chloride films (PVC), fiberglass, acrylic and polycarbonate can be used as greenhouse covering materials.

### Hot Beds

The hotbed is a small, low structure, use for same purposes as greenhouses. Seedling can be started and leafy cuttings rooted in hotbed early in the season. Heat is provided below the propagating medium by electric heating cables, hot waters, steam pipes, hot air fumes or fermenting manures For small propagation operations, hotbed structures are suitable for producing many thousand of nursery plants without the expenditure of large runs for walk-in-green houses.

### Cold Frame

Low-cost cold-frame (sun frame) construction is the same as hotbeds except that no provision is made for supply of bottom heat. A primary use of cold frame is in conditioning or hardening rooted cuttings or young seedlings (linear) preceding field, nursery-row, or container planting. They may be used also for starting new plants in late spring, summer or fall when no external supply of heat is necessary. In cold frame, only the transparent covering is utilized. Close attention to ventilation, shading, watering and winter protection are necessary for success with cold frames.

### Lathhouses

Lathhouses provide outdoor shade and protect container grown plants from high summer temperatures and high light intensities. They reduce moisture stress and decrease the water requirements of plants. Lath houses have many uses in propagation, particularly in conjugation with transplanting and with maintenance of shade-requiring or tender plants. Woven plastic material or polypropylene fabric is widely used in covering structure.

### Propagating Frames

Even in a greenhouse, humidity is not always enough to permit satisfactory rooting of certain kinds of leafy cuttings. Enclosed frames covered with glass or one of the plastic materials may be necessary for successful rooting. There are many variations of such devices (small ones were called wardian cases). Such enclosed frames are also useful for completed graft of small potted nursery stock, since they retain high humidity during the healing process.

In using all such structures, care is necessary to avoid the build up pathogenic organisms. The warm humid conditions combined with lack of air movement and relatively low light intensity, provide excellent conditions for the growth of various fungi and bacteria. Cleanliness of all materials placed in such units is important but in addition, use of fungicide is some time necessary.

## Area of Application

These structures may be applied for off season seedling production (*e.g.*, Papaya, guava, citrus). Asexual propagation (*e.g.*, citrus, mango, guava), winter protection (*e.g.*, papaya, litchi, guava, mango), high temperature protection (*e.g.*, mango, papaya, litchi, guava, citrus).

## Assignments

1. Calculate the off season seedling production of 100 square metres of greenhouse area for papaya, citrus and litchi seedlings saplings?

2. Calculate the high temperature protection of seedlings saplings of mango in the greenhouse measuring 100 square metres.

# Exercise No. 43

## Objective

Practice of *in-vitro* cloning of fruits-preparation of media and sterilization.

## Introduction

The standard formulae for tissue culture media have been determined by research scientists to provide optimum nutrients and growth regulators for specific plants. The formulae developed by Murashige and Skoog (1962) commonly known as MS medium are probably the best known standard formulae and which are used primarily for herbaceous foliage plants. The woody plant medium (WPM) developed by Brant Mc Cown and Grey Lloyd is designed to optimize tissue culture of certain woody plants. Combining the various chemicals for the media in which plant cultures grow is an art, similar to that of cooking a meal-the ingredients and language are different and some of the equipment may not be familiar but recipes usually need to be followed in a precise, orderly manner. The nutrients in media affect the health, vigour and growth of the specialized cells, tissues and organs as the plantlets differentiate and develop in culture there, no single medium can be suggested for all types of plants and organs, so that the details of the culture medium need to be worked out for each plant materials separately.

In general the medium contains:

## Inorganic Salts

These are divided into two groups: major and minor salts.

1. **Major Salts:** The salts of potassium, nitrogen, calcium magnesium, phosphorus and sulphur constitute the major salts. Nitrogen is generally

used as nitrate or ammonium salts, sulphur as sulphates and phosphorus as phosphates.

2. **Minor salts:** The salts of iron, zinc, manganese, boron, copper, cobalt, molybdenum, iodine, etc. make up the minor salts. These salts are essential for the growth of tissues and are required in trace quantities.

### Vitamins

The b-vitamins play an important role in the growth of tissues. Thiamine, nicotinic acid and pyridoxine are generally incorporated in all media, although pantothenic acid, folic acid, biotin, riboflavin, etc. have also been used.

### Growth Regulators

Growth as well as differentiation of tissues *in vitro* is controlled by various growth regulators *i.e.,* auxins, cytokinins, gibberellins, ethylene, abscisic acid etc.

### Carbon Source

Plant tissues in culture can utilize a variety of carbon, carbohydrates-sucrose, glucose, fructose, etc. Sucrose, at 2-5 per cent concentrations in the nutrient media remains the most widely used carbohydrate source.

### Organic Supplements

Complex substances such as yeast-extract malt-extract and casein hydrolysate are also added (0.1-1 per cent w/v). Among various plant extracts, liquid endosperm of immature coconut (coconut water) is widely used (5-20 per cent v/v).

### Media Preparation

Stock solution of major salts, minor salts, iron-EDTA, and vitamins are prepared separately and numbered MS I to V by weighing various chemicals as per specifications and dissolved separately to avoid precipitation. For preparing one litre of the medium: Transfer 50 ml MSI in a beaker and add 20 ml each of MS II and III and MS IV and V each 10 ml. Add 100 mg of inositol and 30 g of sucrose (until otherwise stated). Makeup the volume to one litre. Add the growth regulators from the stock solutions of the growth regulators. Stir the medium thoroughly. Adjust the pH to 5.7 ± 0.1 using a pH meter. Boil the medium and when the temperature of the medium reaches near 70°C, add agar @ 0.8 per cent for solid media. Stire thoroughly and upon thorough mixing of agar, dispense the media in culture vessel and plug the vessel. The media is autoclaved at 121°C or 15 psi for 15 min. The autoclaved media is kept overnight before use.

### Assignments

1. Give the MS media components in detail?

# Exercise No. 44

## Objective

Selection and preparation of explants.

## Introduction

An explants is a piece of plant from which a culture is started. Theoretically, a single explants can produce an infinite number of plants. Explants should always be collected from a healthy disease free mother (stock) plant. Before taking explants, stock plants should be moved to a clear green house, screen house or other shelter away from dust and disease where the mother plant can be observed and given special care for a time, usually about 2 weeks to 6 months. Explants range in size from a microscopic $10^{th}$ of a millimeter to pieces of several centimeters in length. Explants can be meristems, shoot tips, macerated stem pieces, nodes, buds, flowers, peduncle, anthers, petals, leaf discs, petiole, seed nucellus tissue, embryos, seedlings, hypocotyls, bulblets, bulb scales, cormels, radicles, stolen, rhizome tips, root pieces or though rarely single cell or protoplast. Just as juvenility is a factor in selecting material for cuttings, so it is in selecting explants for tissue culture. In general, the juvenile the explants material, the greater the likelihood of success. When shootings, stems, buds, or flowers are taken for explants, they should be cut longer than the final size to be used. In most instances, about 2.5 cm is a convenient size to process.

## Some Points to Remember

- ☆ The small the explants, the less contamination to remove but the larger the explants, the more tissue there is to help establish it in culture.
- ☆ Use new shoots wherever possible because they are cleaner than old wood.
- ☆ Plants under cover or shoots from forced, dorment branches are usually cleaner than field plants.

- ☆ Explants from younger plants often respond more quickly than those taken from older plant.
- ☆ Plant material at the base of a plant is younger than growth high up. Plant material at the base of a plant also can be less clean, due to proximity to the ground.
- ☆ In plastic bags are fastened over actively growing shoot tips on stock plants, the new growth will be cleaner for cutting explants days or weeks later, be sure to provide shade to prevent sunburn.
- ☆ Explants may do better if taken in the morning rather than later in the day.
- ☆ Cut explants with cutters than have been sterilized by dipping in 1/10 bleach and place in a plastic bag containing a moist paper towel and use then a fresh.
- ☆ Any final trimming and treatment should be done in the sterile environment of the transfer hood.

## Explant Cleaning and Treatment

After explants are deteched from the source plant, they must be submitted to cleaning treatments. There are no set rules for cleaning explants because the amount and kind of cleaning depends on how clean or dirty the plant material is to begin with. The process selected also depends on how easily the explants is damaged by a cleaning agent. Disinfectants will not eliminate endogenous contaminants. The most commonly cleaning agents are sodium hypochlorite, mercuric chloride, ethyl alcohol, isopropyl alcohol etc. Use each disinfectant separately as mixing may lead to dangerous reaction. The duration of explants sterilization depends on concentration of sterilent nature of explants (tenderness).

## Assignments

1. How the explants are prepared?

# Exercise No. 45

## Objective

Establishment explant and proliferation.

## Introduction

It is better to start cultures singly in test-tubes instead of in jars, because in this way if one start becomes contaminated it will not contaminate others, as it would be surrounded by others in a jar. The explants collected from the healthy, disease free mother plant should be washed thoroughly with tap water and detergent to remove the traces of dust particles. These explants are then sterilized under the laminar air flow cabinet with surface sterilants to disinfect the explants. The treated explants must be washed with autoclaved distilled water to remove the traces of disinfected as deposition of same may result in deterious effect. The explants are cultured on the medium prepared for the purpose with the help of autoclaved forcep and scalpel. The infection free established culture shows grower symptoms. Accordingly, the culture is excised into two-three segments depending upon the amount of growth and transferred to solid a liquid multiplication containing contamination of cytokinin and auxin. The concentration, combination of growth regulators depends on the type of explants. In some cases clumps formation takes places and in other long growth occurs. The surface sterilized material when inoculated on sterile nutrient media and incubated at 25 ± 1°C with a definite photoperiod and light intensity grows to form a large number of shoots. The chances for successful development of multiples are more when:

(i) The chances for successful development of multiples are more with explants from seedlings or juvenile.

(ii) The size of the explants and the season in which the cultures is established.

(iii) Explants from several woody species secrete a brown exudates which diffuse into the medium and blackens it, the tissue become necrotic and die. This browning can often be minimized by treating the explants with autioxidants like citric acid (0.05 – 0.5 per cent), polyvinyl pyrodiodone (0.1-0.5 per cent), dithiothritol (0.04 per cent) used either in solid or liquid media. Frequent subcutting is another remedy.

**Assignments**

1. What is browning. How it is checked in the medium?

# Exercise No. 46

**Objective**

Stooling in guava.

**Introduction**

In India, guava is propagated by inarching, giving a very high percentage of success during rainy season. But inarching is a cumbersome and labour consuming method and gives limited number of plants from the mother plant in a year. Budding has been adopted only on a limited scale in some parts of the country. Budding is also a cumbersome method. Due to less percentage of success in air layering, only stooling is the alternate method of plant propagation in guava after inarching.

For preparation of guava stool beds, the inarched plants of guava are planted in the month of July at 1.0×1.5m planting distance. These plants are planted in such a way that the union point is remain 5-6 cm below the ground level. These are allowed to grow till February (next year). In February, these are headed back leaving about 10-15 cm portion of stem above the ground level. Regular irrigation is done in the stoolbeds. Weeding is also done from time to time. In the month of March, new shoots are emerged from the beheaded stumps. On an average, 5-6 shoots per stump are emerged. In the month of June, basal portion of shoots are covered with the soil. In the month of August, after removing the soil from the base of the shoots, 2-3 cm wide ring of base is removed from the base of each shoots leaving 2-3 cm portion from the base. Rooting hormones like IBA (8000 ppm) made in lanolin paste is applied to the girdled portion and all the shoots are mounded with the soil to a height of about 25-30 cm. These stooled shoots are left as such for about 2 months. After 2 months, the rooted stooled shoots are detached from mother plants and immediately planted in the nursery and irrigated from time to time. After 4 or 5 months, these plants are ready for sale.

## Assignments

1. Why stooling is done in guava?

# Exercise No. 47

## Objective

Balling and burlapping of nursery plants.

## Introduction

Balling and burlapping (B and B) is also called as Boxing of trees and shrubs. This is an important means of packing of trees and shrubs for market. This techniques is especially useful for large trees and evergreen ones of various sizes. Although some plants can be successfully transplanted without a large ball of earth but balling and burlapping has become an accepted technique to transplant large trees and shrubs. American Stadard for Nursery Stock recommends minimum sizes of balls for four general plant types and these are:

- ☆ Type I Spreading Conifer and Broad-leaved Evergreens.
- ☆ Type II Cone and Broad upright Conifers and Broad-leaved Evergreens.
- ☆ Type III Columnar Evergreens.
- ☆ Type IV Standard Shade Trees.

American Standard for Nursery Stock recommends their suggestion for those plants that have been field grown under favourable conditions. Variation in root system of plants, soil conditions and cultural practices also affect the ball size but it also depends on the integrity of the nurseryman. Size of ball varies from plant to plant *e.g.*, For Type I plants ball diameter varies from 0.25 m to 0.90 m, for Type II plants it varies from 0.25 m to 1.25 m, similarly for Type III plants, it varies from 0.25 m to 1.00 m and for Type IV plants, it varies from 0.45 m to 2.00 m. To assure successful transplanting, ball size should be of a diameter and depth, large enough to obtain the fibrous roots, which are most important plant part for transplantation.

Recommended Balling and Burlapping specifications as per American Association of Nurserymen (1990) is given below:

**Table 1: Spreading Conifer and Broad-Leaved Evergreens (Type I)**

| *Spread (m)* | *Diameter (m)* |
|---|---|
| 0.45 | 0.25 |
| 0.60 | 0.30 |
| 0.75 | 0.35 |
| 1.05 | 0.45 |
| 1.20 | 0.52 |
| 1.50 | 0.60 |
| 1.80 | 0.70 |
| 2.10 | 0.80 |
| 2.40 | 0.90 |

**Table 2: Cone and Broad Upright Conifers and Broad-Leaved Evergreens (Type II)**

| *Height (m)* | *Diameter (m)* |
|---|---|
| 0.45 | 0.25 |
| 0.60 | 0.30 |
| 0.90 | 0.35 |
| 1.50 | 0.50 |
| 1.80 | 0.55 |
| 2.10 | 0.60 |
| 2.40 | 0.67 |
| 2.70 | 0.75 |
| 3.00 | 0.85 |
| 3.60 | 0.95 |
| 4.20 | 1.05 |
| 4.80 | 1.15 |
| 5.40 | 1.25 |

**Table 3: Columnare (Type III)**

| *Height (m)* | *Diameter (m)* |
|---|---|
| 0.45 | 0.25 |
| 0.60 | 0.30 |
| 0.90 | 0.32 |
| 1.20 | 0.35 |
| 1.50 | 0.40 |

*Contd...*

**Table 3–*Contd...***

| *Height (m)* | *Diameter (m)* |
|---|---|
| 1.80 | 0.45 |
| 2.10 | 0.50 |
| 2.40 | 0.55 |
| 2.70 | 0.60 |
| 3.00 | 0.67 |
| 3.60 | 0.75 |
| 4.20 | 0.82 |
| 4.80 | 0.90 |
| 5.40 | 1.00 |

**Table 4: Standard Shade Trees (Type IV)**

| *Calliper (cm)* | *Diameter (m)* |
|---|---|
| 3.12 | 0.45 |
| 3.33 | 0.50 |
| 4.37 | 0.55 |
| 5.00 | 0.60 |
| 6.25 | 0.70 |
| 7.50 | 0.80 |
| 8.75 | 0.95 |
| 10.00 | 1.05 |
| 11.25 | 1.20 |
| 12.50 | 1.35 |
| 15.00 | 1.50 |
| 17.50 | 1.75 |
| 20.00 | 2.00 |

Ball depth ratio reflect root distribution which will vary with different tree species, the region of the country where they are grown and the soil type, whereas, for a successful transplantation of large trees and shrubs by balling and burlapping techniques many diameter/depth ratios were recommended by American Standard for Nursery Stock.

## Assignments

1. If a standard shade tree having caliper size is 5 cm. what will be diameter (m) of the tree?

# Appendices

# Appendix I

## List of Fruit Plants

| *Common Name* | *Botanical Name* | *Family* |
|---|---|---|
| Almond | *Prunus amygdalus Batsch* | Rosaceae |
| Apple | *Malus pumila* Mill. | Rosaceae |
| Aonla | *Emblica officinalis* L. | Euphorbiaceae |
| Apricot | *Prunus armeniaca* L. | Rosaceae |
| Atemoya | *Annona atemoya* Hort. | Annonaceae |
| Avocado | *Persea americana* Mill. | Lauraceae |
| Bael | *Aegle marmelos* Correa | Rutaceae |
| Banana | *Musa paradisiaca* L. | Musaceae |
| Indian Ber | *Ziziphus mauritiana* Lamk. | Rhamnaceae |
| Chinese Ber | *Ziziphus jujube* Mill. | Rhamnaceae |
| Bramble | *Rubus spp.* | Rosaceae |
| Bullocks heart | *Annona reticulata* | Annonaceae |
| Cape gooseberry | *Physalis peruviana* L. | Solanaceae |
| Carambola | *Averrhoa carambola* L. | Oxallidiaceae |
| Karonda | *Carissa cadandas* L. | Apocynaceae |
| Cashew nut | *Anacardium occidentale* L. | Anacardiaceae |

| *Common Name* | *Botanical Name* | *Family* |
|---|---|---|
| Cherry (sweet) | *Prunus avium* L. | Rosaceae |
| Cherry (sour) | *Prunus cerasus* L. | Rosaceae |
| Chestnut | *Castanea sativa* Mill. | Fagaceae |
| Citron | *Citrus medica* L. | Rutaceae |
| Grapefruit | *Citrus paradisi* Macf. | Rutaceae |
| Kumquat | *Fortunella japonica* Swingle | Rutaceae |
| Lemon | *Citrus limon* Burm. | Rutaceae |
| Lime | *Citrus aurantifolia* Swingle | Rutaceae |
| Mandarin | *Citrus reticulata* Blanco | Rutaceae |
| Pummelo | *Citrus grandis* Merr. | Rutaceae |
| Karna Khatta | *Citrus karna* Raf. | Rutaceae |
| Sapota | *Achras sapota* L. | Sapotaceae |
| Sour orange | *Citrus aurantium* L. | Rutaceae |
| Sweet lime | *Citrus limettioides* Tanaka | Rutaceae |
| Sweet orange | *Citrus sinensis* Osbeck | Rutaceae |
| Trifoliate orange | *Poncirus trifoliata* Raf. | Rutaceae |
| Crab apple | *Malus baccata* Borkh. | Rosaceae |
| Date palm | *Phoenix dactylifera* L. | Palmae |
| | *Phoenix sylvestris* Roxb. | Palmae |
| Fig | *Ficus carica* L. | Moraceae |
| Grapes | *Vitis vinifera* L. | Vitaceae |
| Guava | *Psidium guajava* L. | Myrtaceae |
| Hog plum | *Spondias cytherea* Sonn. | Anacardiaceae |
| | *Spondias pinnate* Kur. | Anacardiaceae |
| Indian almond | *Terminalia catappa* L. | Combretaceae |
| Jackfruit | *Artocarpus heterophylus* Lam. | Moraceae |
| Jambolan (Jamun) | *Syzygium cuminii* Skeels | Myrtaceae |
| Jamua | *Syzygium fruiticosum* Roxb. | Myrtaceae |
| Khirni | *Manilkara hexandra* Dubard. | Sapotaceae |
| Litchi | *Litchi chinensis* Sonn. | Sapindaceae |
| Loquat | *Eribotrya japonica* Lindl. | Rosaceae |
| Mango | *Mangifera indica* L. | Anacardiaceae |

| *Common Name* | *Botanical Name* | *Family* |
|---|---|---|
| Mahua | *Basia latifolia* Macbride | Sapotaceae |
| Monkey Jack (Barhal) | *Artocarpus lakoocha* Roxb. | Moraceae |
| Mulberry | *Morus alba* L. | Moraceae |
| Netal plum | *Carissa grandiflora* | Apocynaceae |
| Papaya | *Carica papaya* L. | Caricaceae |
| Peach | *Prunus persica* Batsch. | Rosaceae |
| Pear | *Pyrus communis* L. | Rosaceae |
| Pecan | *Carya illinoensis* Koch. | Juglandaceae |
| Persimmon | *Diosypyros kaki* L. | Ebenaceae |
| Phalsa | *Grewia subinalqualis D.C.* | Tilliaceae |
| Walnut | *Juglans regia* L. | Jaglandaceae |
| Pineapple | *Ananas comosus* Mur. | Bromeliaceae |
| Pistachio Nut | *Pistacia vera* L. | Rosaceae |
| Plum | *Prunus domestica* | Rosaceae |
| | *Prunus salicina* Lindl. | Rosaceae |
| Pomegranate | *Punica granatum* L. | Punicaceae |
| Quince | *Cydonia oblonga* Mill. | Rosaceae |
| Rambutan | *Nephelium lappaceum* L. | Sapindaceae |
| Roseapple | *Syzygium jambos* L. | Myrtaceae |
| Soursop | *Annona muricata* L. | Annonaceae |
| Strawberry | *Fragaria vesca* L. | Rosaceae |
| Strawberry guava | *Psidium cattleianum* | Myrtaceae |
| Custard apple | *Annona squamosa* L. | Annonaceae |
| Tamarind | *Tamarindus indica* L. | Leguminosae |
| Wood apple (Kaitha) | *Limonia acidissima* L. | Rutaceae |

# Appendix II

## List of Vegetable Crops

| *Common Name* | *Botanical Name* | *Family* |
|---|---|---|
| Onion | *Allium cepa* | Amaryllidaceae |
| Leek | *Allium porrum* | Amaryllidaceae |
| Garlic | *Allium sativum* | Amaryllidaceae |
| Chive | *Allium schoenoprassum* | Amaryllidaceae |
| Colocasia | *Colocasia esculenta* | Araceae |
| Alocacia | *Alocasia* sp. | Araceae |
| Yam | *Dioscorea alata* | Dioscoreaceae |
| Suthani | *Discorea esculenta* | Dioscoreaceae |
| Asparagus | *Asparagus officinalis* | Liliaceae |
| Popcorn | *Zea mays* var. evarta | Graminae |
| Sweet corn | *Zeamays* var. rugosa | Graminae |
| Lettuce (Head) | *Lactuca sativa* Var. capitata | Aesteraceae |
| Lettuce (Leaf) | *Lactuca sativa* | Aesteraceae |
| Newzealand spinach | *Tetragonia expansa* | Aizoaceae |
| Celery | *Apium graveolens* | Apiaceae |
| Coriander | *Coriandrum stivum* | Apiaceae |
| Carrot | *Daucus carota* | Apiaceae |

| *Common Name* | *Botanical Name* | *Family* |
|---|---|---|
| Parsnip | *Partinaea sativa* | Apiaceae |
| Amaranth | *Amaranthus candatus* | Amaranthaceae |
| Poi(green) | *Basella alba* | Basellaceae |
| Poi(violet) | *Basella rubra* | Basellaceae |
| Horse radish | *Armoracia rusticana* | Brassicaceae |
| Turnip | *Brassica compestris* var. rapa | Brassicaceae |
| Chinese cabbage | *Brassica chinensis* | Brassicaceae |
| Kale | *Brassica oleracea* var. acephala | Brassicaceae |
| Cauliflower | *Brassica oleracea* var. botrytis | Brassicaceae |
| Cabbage | *Brassica oleracea* var. capitata | Brassicaceae |
| Brussels sprout | *Brassica oleracea* var. gemmifera | Brassicaceae |
| Knol khol | *Brassica oleracea* var. gongylodes | Brassicaceae |
| Broccoli | *Brassica oleracea* var. italica | Brassicaceae |
| Mustard | *Brassica juncea* | Brassicaceae |
| Rutabag | *Brassica napus* var. nabobrassica | Brassicaceae |
| Chinese cabbage | *Brassica pekenensis* | Brassicaceae |
| Chinese radish | *Raphanus sativus* var. longipinnatus | Brassicaceae |
| Beet | *Beta vulgaris* | Chinopodiaceae |
| Beet leaf (Palak) | *Beta vulgaris* var. bengalensis | Chinopodiaceae |
| Chenopodium | *Chenopodium alba* | Chinopodiaceae |
| Chicary | *Cichorium intypus* | Compositae |
| Endive | *Cichorium andivia* | Compositae |
| Jerusallam artichoke | *Helianthus tuberosum* | Compositae |
| Sweet potato | *Ipomea batata* | Convolvulaceae |
| Cress | *Lapidium sativum* | Cruciferae |
| Ashgourd | *Benincasa hispida* | Cucurbitaceae |
| Watermelon | *Citrullus lanatus* | Cucurbitaceae |
| Kundru or Ivory | *Coccinia indica* | Cucurbitaceae |
| Muskmelon | *Cucumis melo* var. cantalupensis | Cucurbitaceae |
| Long melon | *Cucumis melo* var. utilissimus | Cucurbitaceae |
| Cucumber | *Cucumis sativus* | Cucurbitaceae |

| *Common Name* | *Botanical Name* | *Family* |
|---|---|---|
| Winter squash | *Curcurbita maxima* | Cucurbitaceae |
| Pumpkin | *Curcurbita moschata* | Cucurbitaceae |
| Summer squash | *Cucurbita pepo* | Cucurbitanceae |
| Bottlegourd | *Lagenaria sicigerania* | Cucurbitaceae |
| Ridgegourd | *Luffa acuitengula* | Cucurbitaceae |
| Smoothgourd | *Luffa aestica* | Cucurbitaceae |
| Spongegourd | *Luffa cylindrical* | Cucurbitaceae |
| Bittergourd | *Momordica charantia* | Cucurbitaceae |
| Snakegourd | *Trichosanthus anguina* | Cucurbitaceae |
| Pointedgourd | *Trichosanthus dioica* | Cucurbitaceae |
| Tapioca | *Manihot esculenta* | Euphorbiaceae |
| Swordbean | *Canavalia ensiformis* | Fabaceae |
| Bean | *Dolichos lablab* | Fabaceae |
| Limabean | *Phaseolus lunalus* | Fabaceae |
| Frenchbean | *Phaseolus vulgaris* | Fabaceae |
| Pea | *Pisum arvensis* | Fabaceae |
| Garden pea | *Pisum sativum* | Fabaceae |
| Wingedbean | *Psophocarpus tetragonolobus* | Fabaceae |
| Kasuri Methi | *Trigonella corniculata* | Fabaceae |
| Broadbean | *Vicia faba* | Fabaceae |
| Cowpea | *Vigna unguiculata* | Fabaceae |
| Fenugreek | *Trigonella foenumgraecum* | Fabaceae |
| Mint | *Mentha pipereta* | Lebiateae |
| Okra | *Abelmoschus esculentus* | Malvaceae |
| Roselle | *Hibiscus subdariffa* | Malvaceae |
| Black pepper | *Piper nigrum* | Piperaceae |
| Kulpha | *Portulaca oleracea* | Portulaceae |
| Hot pepper | *Capsicum annum* | Solanaceae |
| Sweet pepper | *Capsicum fruitescens* | Solanaceae |
| Tomato | *Lycopersicon esculenium* | Solanaceae |
| Brinjal | *Solanum melongena* | Solanaceae |
| Potato | *Solanum tuberosum* | Solanaceae |

# Appendix III

**List of Medicinal and Aromatic Plants**

| *Common Name* | *Botanical Name* | *Family* |
|---|---|---|
| Kapoor Tulsi | *Ocimum americanum* | Labiatae |
| Tulshi | *Ocimum sanctum* | Labiatae |
| Bergamont mint | *Mentha citrate* | Labiatae |
| Japanese mint | *Mentha arvensis* | Labiatae |
| Pepper mint | *Mentha piperita* | Labiatae |
| Spear mint | *Mentha spicata* | Labiatae |
| Kujai | *Rosa moschata* | Rosaceae |
| Ghingaru | *Crataegus oxyacantha* | Rosaceae |
| Kateri | *Solanum khasianum* | Solanaceae |
| Khurasani Ajvayan | *Hyocymus albus* | Solanaceae |
| Keu | *Costus speciosus* | Zingiberaceae |
| Ilaychi | *Amomum zanthioides* | Zingiberaceae |
| Sarpgandha | *Rauwolfia serpentine* | Apocynaceae |
| Sadabhar | *Catheranthus* sp. | Apocynaceae |
| Bahera | *Terminalia velerica* | Combretacaceae |
| Kantala | *Agave americana* | Amaryllidaceae |

| *Common Name* | *Botanical Name* | *Family* |
|---|---|---|
| Adulasa | *Adhatoda vasica* | Acanthaceae |
| Bach | *Acorus calamus* | Araceae |
| Ghikanvar | *Aloe vera* | Liliaceae |
| Satavar | *Aspargus recemosus* | Liliaceae |
| Aonla | *Emblica offacinalis* | Euphorbiaceae |
| Brahmi | *Bacopa monnieri* | Scrophulariaceae |
| Gulvanphasa | *Viola serpens* | Violaceae |
| Ratalu | *Dioscorea deltoidae* | Dioscoreaceae |
| Buckwheat | *Fagopyrum esculentum* | Polygonaceae |
| Antmul | *Tylophora indica* | Aslepiadaceae |
| Hathi Bach | *Sansveria* Sp. | Haemodoraceae |
| Isabgol | *Plantago ovate* | Plantaginaceae |
| Piplamul | *Piper longum* | Piperaceae |
| Khas, vetiver | *Vetiveria zizanioides* | Gramineae |
| Citronella java | *Cymbopogon winterianus* | Gramineae |
| Lemongrass | *Cymbopogon flexuosus* | Gramineae |
| Palmarosa | *Cymbopogon martini* | Gramineae |

# Appendix IV

## List of Spices and Condiments

| *Common Name* | *Botanical Name* | *Family* |
|---|---|---|
| Ajowan | *Trachyspermum ammi* L. | Umbelliferae |
| Allspice or pimento | *Pimenta officinalis* L. | Myrtaceae |
| Angelica | *Angelica archangelica* L. *(Syn. Archangelica officinalis,* Hoffmann*)* | Umbelliferae |
| Aniseed | *Pimpinella anisum* L. *(Syn. Anisum vulgare,* Gaertner*) Anisum officinalis,* Moench. | Umbelliferae |
| Asafoetida | *Ferula asafetida* L. | Umbelliferae |
| Balm or Lemon balm | *Melissa officinalis* L. | Labiatae |
| Basil or Sweet basil | *Ocimum basilicum* L. | Labiatae |
| Bay or Laurel leaves | *Laurus nobilis* L. | Lauraceae |
| Caper | *Capparis spinosa* L. | Capparidaceae |
| Capsicum or chillies | *Capsicum annum* L. | Solanaceae |
| Caraway | *Carum carvi* L. | Umbelliferae |
| Large cardamom | *Aframomum* Sp. | Zingiberaceae |
| Greater cardamom | *Amomum* sp. | Zingiberaceae |

| *Common Name* | *Botanical Name* | *Family* |
|---|---|---|
| Lesser cardamom | *Elettaria cardamomum* | Zingiberaceae |
| Cassia | *Cinnamomum aromaticum* | Lauraceae |
| Cinnamom | *Cinnamomum zeylanicum* | Lauraceae |
| Celery seed | *Apium graveolens* Line Var. *Chanu elery* | Umbelliferae |
| Celeriac | *Apium graveolens* Var. rapaccum | Umbelliferae |
| Chervil | *Anthriscus cerefolium* Hoffn. | Umbelliferae |
| Chives or Cives | *Allium schoenoparasum* L. | Liliaceae |
| Clove | *Eugenia caryophyllus* *(Syn. Syzygium aromaticum)* | Myrtaceae |
| Coriander | *Coriandrum sativum* L. | Umbelliferae |
| Cumin seed | *Cuminum cyminum* L. | Umbelliferae |
| Cumin black | *Nigella sativa* L. | Umbelliferae |
| Curry leaf | *Murraya koenigii* L. | Rutaceae |
| Dill and Indian dill sowa | *Anethum sowa (sowa)* | Umbelliferae |
| Fennel | *Foeniculum vulgare* *(Syn. F.officinale)* | Umbelliferae |
| Fenugreek | *Trigonella foenum graecum* | Leguminosae |
| Galangal | *Alpinia galangal* | Zingeberaceae |
| Garlic | *Allium sativum* | Lilliaceae |
| Ginger | *Zingiber officinale* | Zingiberaceae |
| Horse raddish | *Cochlearia armoracia* | Cruciferae |
| Hyssop | *Hysssopus officinalis* | Labiatae |
| Juniper | *Juniperous communis* | Pinaceae |
| Kokam | *Garcinia indica* | Guttiferae |
| Stone leek | *Allium fistulosum* | Lilliacae |
| Lovage | *Levisticum officinale* | Umbelliferae |
| Mace | *Myristica fragrans* | Myristicaceae |
| Marjoram | *Marjorana fragrans* | Labiatae |
| Marjoram | *Marjorma hortensis* | Labiatae |
| Mint or Japanese mint | *Mentha arvensis* | Labiatae |

| *Common Name* | *Botanical Name* | *Family* |
|---|---|---|
| Mustard, True or Black | *Brassica nigra* | Brassicaceae |
| Safed rai or white mustard | *Brassica alba* | Brassicaceae |
| Indian or brown mustard | *Brassica juncea* | Brassicaceae |
| Nutmeg | *Myristica fragrans* | Myristicaceae |
| Onion | *Allium cepa* | Liliaceae |
| Oregano or Origanum | *Origanum vulgare* | Labiatae |
| Parsley | *Petroselinum crispum* | Umbelliferae |
| Black pepper | *Piper nigrum* | Piperaceae |
| Pepper long | *Piper longum* | Piperaceae |
| Peppermint | *Mentha pipereta* | Labiatae |
| Poppy seed | *Papaver somniferum* | Papaveraceae |
| Rose mary | *Rosmarinus officinalis* | Labiatae |
| Saffron | *Crocus sativus* | Iridaceae |
| Sage | *Salvia officinalis* | Labiatae |
| Savory | *Satureia hortensis* | Labiatae |
| Shallot | *Allium ascalonicum* | Labiatae |
| Spear mint | *Mentha spicata* | Labiatae |
| Star-anise | *Illiticium verum* | Magnoliaceae |
| Sweet flag or Calamus | *Acorus calamus* | Araceae |
| Tamarind | *Tamarindus indica* | Leguminosae |
| Tarragon | *Artimisia dracunculus* | Compositae |
| Thyme | *Thymus serpyllum* | Labiatae |
| Turmeric | *Curcuma longa* | Zingiberaceae |
| Vanilla | *Vanilla fragrans* | Orchidaceae |

# Appendix V

## List of Plantation Crops

| *Common Name* | *Botanical Name* | *Family* |
|---|---|---|
| Tea | *Camellia sinensis* | Ternstroemiaceae |
| Coffee | *Coffea arabica* | Rubiaceae |
| Coffee | *Coffea canephora* | Rubiaceae |
| Cacao | *Theobroma cacao* | Sterculiaceae |
| Rubber | *Hevea brasiliensis* | Euphorbiaceae |
| Coconut | *Cocos nucifera* L. | Palmae |
| Arecanut or Betel nut | *Areca catechu* | Palmae |
| Plantain | *Musa paradisiaca* | Musaceae |

# Appendix VI

## List of Flowering Annuals

| *Common Name* | *Botanical Name* | *Family* | *Flower Colour* |
|---|---|---|---|
| Sand-verbena | *Abronia umbellate* | Nyctaginaceae | Rosy lilac |
| Acroclinium | *Acroclinium roseum* | Compositae | White, pink |
| Pheasant's etye cup like | *Adonis aestivalis* | Ranunculaceae | Dark crimoson, butter |
| Floss flower | *Ageratum* spp. | Compositae | Pale lavender to deep mauve, blue, white |
| Corn Cockle | *Agrostemma githago* | Caryophyllaceae | Pale lilac |
| Hollyhock | *Althaea rosea* | Malvaceae | White to cream, yellow |
| Sweet alyssum | *Alyssum maritimum* | Cruciferae | White, lilac, rose-pink |
| Amaranth | *Amaranthus caudatus* | Amaranthaceae | White, palegreen, crimson |
| Alkanet, Bugloss | *Anchusa capensis* | Boraginaceae | Deep blue with white centre |
| Antirrsinum/ snap dragon | *Antirrhimum majus* | Scrophulariaceae | White, yellowm, pink, rose, mauve |
| African daisy | *Arctotis grandis* | Compositae | White with litac centre |
| Blazing star | *Bartonia aurea* | Loasaceae | Orange colour |
| Daisy or English Daisy | *Bellis perennis* | Compositae | White, pink, crimson |

| *Common Name* | *Botanical Name* | *Family* | *Flower Colour* |
|---|---|---|---|
| Swan river daisy | *Brachycome iberidifolia* | Compositae | White, blue, pale |
| Browallia | *Browallia demissa* | Sonanaceae | Blue, white |
| Tassel flower | *Cacalia coccinea* | Compositae | Orange, dull |
| Slepperwort/ pouch flower | *Calceolaria spp.* | Scorphulariaceae | Bright coloured |
| Calendula | *Calendual officinalis* | Compositae | Straw colour to deep orange |
| Calliopsis, Tick seed | *Coreopsis tinctoria* | Compositae | Yellow, creamson brown |
| Cosmea, cosmos | *Cosmos bipinnatus* | Compositae | Rose, pink, purple, white, yellow |
| Cigar plant | *Cuphea ignea* | Lythraceae | Purple base white mouth, scarlet |
| Hound's tongue, Chinese forgetmenot | *Cynoglossum amabile* | Boraginaceae | White, blue |
| Dahlia | *Dahlia* spp. | Compositae | Vary colour with disc |
| Larkspur | *Delphinium ajacis* | Ranunculaceae | White, blue, purple |
| Sweet william | *Dianthus barbatus* | Caryophyllaceae | White, blue, purple |
| Carnation | *Dianthus caryophyllus* | Caryophyllaceae | White, pink, crimson, mauve, yellow, violet, red |
| Common foxglove | *Digitalis purpurea* | Scrophulariaceae | White, apricot, pink, cream |
| Cape daisy, African daisy | *Dimorphotheca sinuate* | Compositae | White, bluish white, yellow |
| Viper's bugloss | *Echium plantagineum* | Boraginaceae | Blue, mauve, purple, rose |
| Californian poppy | *Eschscholzia californica* | Papaveraceae | Yellow, Lemon, creamy white |
| Spurge | *Euphorbia heterophylla* | Euphrbiaceae | Orange red |
| Kingfisher daisy | *Felicia bergeriana* | Compositae | Metallic blue flowers |
| Blanket flower | *Gaillardia pulchella* | Compositae | Yellow, Lemon, orange |
| Treasure flower | *Gazania splendens* | Compositae | Red, pink, orange, yellow |

| *Common Name* | *Botanical Name* | *Family* | *Flower Colour* |
|---|---|---|---|
| Gilia | *Gilia capitata* | Polemoniaceae | Yellow, red, orangem purple |
| Globe amaranth | *Gomphrena globosa* | Amaranthaceae | Purple, violet, rose pink |
| Baby's Breath | *Gypsophila elegans* | Caryophyllaceae | White, pink, yellow |
| Sun flower | *Helianthus annuus* | Compositae | Yellow |
| Straw flower | *Helichrysum bracteatum* | Compositae | Yellow, pink, red |
| Heliotrope/ Cherry Pie | *Heliotropium arborescens* | Boraginaceae | White to blue, purple |
| Candytuft | *Iberis amara* | Cruciferae | White, pink, purple, litac |
| Balsam | *Impatiens balsamina* | Balsaminaceae | Rose, pink, red, violet |
| Sweet pea | *Lathyrus odoratus* | Leguminoseae | White to any colour |
| Sea lavender/ Statice | *Limonium sinuatum* | Plumbaginaceae | White, rose, dark blue |
| Toad flax | *Linaria maroccana* | Scrophulariaceae | White, yellow, pink |
| Scarlet flax | *Linum grandiflorum* | Linaceae | Blue flowers |
| Lobelia | *Lobelia erinus* | Campanulaceae | Blue, crimsom, purple |
| Lupin | *Lupinus mutabilis* | Leguminosae | Dark, blue, white, pink |
| Feverfew | *Matricaria eximia* | Compositae | Ivory white, yellow |
| Gilliflower/Stock | *Matthiola incana* | Cruciferae | White, red, pink, blue |
| Monkey flower | *Mimulus tigrinus* | Scrophulariaceae | Pink, yellow, crimson |
| Four O'clock | *Mirabilis jalapa* | Nyctaginaceae | White, red, yellow |
| Forget me not | *Myosotis alpestris* | Boraginaceae | White, dark blue, pink |
| Nemesia | *Nemesia strumosa* | Scrophulariaceae | White, blue, purple, pure white |
| Baby blue eyes | *Nemophila menzesii insignis* | Hydrophyllaceae | Sky blue, purple, pure white |
| Cup flower | *Nierembergia caerulea* | Solanaceae | Voilet blue, with yellow eye |
| Evening primrose | *Oenothera biennis* | Onagraceae | Yellow, tubular |
| Corn poppy | *Papaver rhoeas* | Papaveraceae | White, pink, blue |
| Petunia | *Petunia hybrid* | Solanaceae | Pink, red, pale blue |
| Phacelia | *Phacelia campanularia* | Hydrophyllaceae | Blue |

| *Common Name* | *Botanical Name* | *Family* | *Flower Colour* |
|---|---|---|---|
| Annual phlox | *Phlox drummondii* | Polemoniaceae | White, pink, purple |
| Lady's lace | *Pimpenella monoica* | Umbelliferae | White |
| Portulaca | *Portulaca grandiflora* | Portulaceae | White, purple, pink, orange, red |
| Cone flower | *Rudbeckia bicolor* | Compositae | Yellow, dark brown |
| Painted tongue | *Salpiglosis sinulata* | Solanaceae | Pink, gold, blue |
| Sage | *Salvia patens* | Labiatae | Blue, white |
| Soapwart | *Saponaria calabrica* | Caryophyllaceae | Pink, white |
| Cineraria | *Senicio cruentus* | Compositae | White, pink blue, purple |
| Catchfly | *Selene pendula* | Caryophyllaceae | Rose, pink, purple |
| Lace flower | *Trachymene* | Umbelliferae | White, pink, blue |
| Garden nasturtium | *Tropaeolum majus* | Tropaeolaceae | Orange, primrose, salmon |
| Namaqualand daisy | *Vanidium fastuosum* | Compositae | Deep orange |
| Verbena | *Verbena spp.* | Verbenaceae | White, pink, purple |
| Pansy | *Viola tricolor hortensis* | Violaceae | Butterfly like flowers yellow, blue, red |
| Viola/Tufted Pansy | *Viola cornuta* | Violaceae | White, blue, yellow |
| Zinnia | *Zinnia elegans* | Compositae | White, cream, crimson |
| Persian carpet | *Zinnia haageana* | Compositae | Wide range of colour |

# Appendix VII

## List of Climbers

| *Common Name* | *Botanical Name* | *Family* | *Flower Colour* |
|---|---|---|---|
| Moonflower | *Calonyction aculeatum* | Convolvulaceae | White, purple |
| Mussel shell Creeper/Butterfly Pea | *Clitoria ternatea* | Leguminosae | White, blue, purple |
| Cobaea | *Cobaea scandens* | Polemoniaceae | Voilet, greenish |
| Morning glory | *Pharbitis purpurea* | Confolvulaceae | White, blue, purple |
| Star ipomoea | *Quamoclit pennata* | Convolvulaceae | White |
| Black eyed susan | *Thunbergia alata* | Acanthaceae | White |
| Canary creeper | *Tropaeolum peregrinum* | Tropaeolaceae | Golden yellow |

# Appendix VIII

## List of Herbaceous Perennials

| *Common Name* | *Botanical Name* | *Family* | *Flower Colour* |
|---|---|---|---|
| Michaelmas daisy/ Perennial Aster | *Aster amellus* | Compositae | White, deep blue, lilac and pink |
| Coreopsis | *Coreopsis grandiflora* | Compositae | Bright, yellow flower |
| Touch-me-not/ Perennial Balsam | *Impatiens hookeriana* | Balsaminaceae | White with purple markings |
| Geranium | *Pelargonium zonale* | Geraniaceae | White, pink, scarlet and crimpson |
| Border phlox | *Phlox decussate* | Polemoniaceae | White, pink, violet, purple, crimson and red |
| Chinese or Japanese Bell flower | *Platycodon grandiflorum* | Campanulaceae | Blue or white |
| Perennial portulaca | *Portulaca grandiflora* | Portulaceae | Many colours, white, pink, red etc. |
| Usambara violet African violet | *Saintipaulia ionantha* | Gesneriaceae | Violet |
| Golden rod | *Solidago canadensis* | Compositae | Golden-yellow |

| *Common Name* | *Botanical Name* | *Family* | *Flower Colour* |
|---|---|---|---|
| Moss verbena | *Verbena erinoides* | Verbenaceae | Pink and white |
| Periwinkle | *Vinca rosea/ Catheranthus* | Apocynaceae | Rosy purple |
| | *Vinca alba* | Apocynaceae | White |
| Sweet violet | *Viola odorata* | Violaceae | White, blue, purple |
| Perennial zinnia | *Zinnia linearis* | Compositae | Golden-yellow |
| | | **Hills** | |
| Yarrow or Milfoil | *Achillea millefolium* | Compositae | White, red or purple |
| Columbine | *Aquilegia vulgaris* | Ranunculaceae | White, red, yellow, pink crimson, blue etc. |
| Elephant ear | *Bergenia cordifolia* | Saxifragaceae | Pink, white |
| Bell flower | *Campanula burghaltii* | Companulaceae | Grey-blue |
| Moon dairy | *Chrysanthemum maximum* | Compositae | White, with yellow central disc |
| Perennial delphimium | *Delphinium hybridum* | Ranumculaceae | Many flowers white, blue sky, blue purple red, scarlet, yellow etc. |
| Common foxglove | *Digitalis purpurea* | Scrophulariaceae | White, pink, crimpson |
| Treasure flowers | *Gazania splendens* | Compositae | Red, pink, orange, yellow etc. |
| Baby breath | *Gypsophila elegans* | Caryophyllaceae | White or pink |
| Coral bells | *Heuchera sanguniea* | Saxifragaceae | White, rose red |
| Maxican tulip poppy | *Hunnemannia fumareifolia* | Papaveraceae | Yellow |
| Peony or Paeony | *Paeonia* spp. | Ranunculaceae | |
| Lupin | *Lupinus hartwegii* | Leguminosae | White, pink, pale blue |
| Bread tongue | *Penstemon barbatus* | Scrophulariaceae | White, pink, purple |
| Baby primrose | *Primula malacoides* | Primulaceae | White, pale, pink, rose red, crimpson |
| Pyrethrum of gardens | *Pyrethrum roseum* | Compositae | White, lilac, rose and red |
| Coneflower | *Rudbeckia bicolor* | Compositae | Yellow with dark brown central disc |
| Cape primrose | *Streptocarpus spp.* | Gesneriaceae | Blue, lilac, mauve, pink, white |
| With bone flower | *Torenia fournieri* | Scrophulariaceae | Violet, pale blue |

# Appendix IX

**List of Important Garden Shrubs**

| *Common Name* | *Botanical Name* | *Family* |
|---|---|---|
| Acalypha | *Acalypha godseffiana* | Euphorbiaceae |
| | *Acalypha hispida* | Euphorbiaceae |
| Allamanda | *Allamanda cathartica* | Apocynaceae |
| Angelonia | *Angelonia grandiflora* | Scrophulariaceae |
| Aralia | *Aralia balfouriana* | Araliaceae |
| Azalea | *Rhododendron indicum* | Ericacea |
| Baleria | *Barleria cristata* | Acanthaceae |
| Orchid tree | *Bauhinia acuminate* | Leguminosae |
| Berberis | *Berberis asiatica* | Berberidaceae |
| Bougainvillea | *Bougainvillea spectabilis* | Nycataginaceae |
| | *Bougainvillea peruviana* | Nycataginaceae |
| | *Bougainvillea glabra* | Nycataginaceae |
| Brunfelsia | *Brunfelsia americana* | Solanaceae |
| Buddleia | *Buddleia asiatica* | Loganiaceae |
| Krishnachura (Bengali) | *Caesalpinia pulcherrima* | Leguminosae |
| Radha chura (Bengali) | *Caesal pinia pulcherrima var. flava* | Leguminosae |

| *Common Name* | *Botanical Name* | *Family* | *Flower Colour* |
|---|---|---|---|
| Bottle brush | *Callistemon lanceolatus* | Myrtaceae | |
| Camellia | *Camellia japonica* | Theaceae | |
| Cassia | *Cassia glauca* | Leguminosae | |
| Lily thorn | *Catesbaea spinosa* | Rubiaceae | |
| Raat ki rani | *Cestrum nocturnum* | Solanaceae | |
| Din ka raja | *Cestrum diurnum* | Solanaceae | |
| Forest jasmine | *Clerodendron inerme* | Verbenaceae | |
| Tubeflower | *Clerodendron siphonanthus* | Verbenaceae | |
| Bengal coffee | *Coffea bengalansis* | Rubiaceae | |
| Scarlet plum | *Euphorbia fulgens* | Euphorbiaceae | |
| Kumquat | *Fortunella japonica* | Rutaceae | |
| Garedenia/Cape Jasmine | *Gardenia lucida* | Rubiaceae | |
| Hamelia | *Hamelia patens* | Rubiaceae | |
| Thalkamal | *Hibiscus mutabilis* | Malvaceae | |
| Gurhal/China rose | *Hibiscus rosesinensis* | Malvaceae | |
| Rangan | *Ixora acuminate* | Rubiaceae | |
| Torch tree | *Ixora, parvilora* | Rubiaceae | |
| Jasmine | *Jasminum grandiflorum* | Oleaceae | |
| Bherenda | *Jatropha multifida* | Euphoriaceae | |
| Guatemala rhubarb | *Jatropha podagrica* | Euphoriaceae | |
| Juniperus | *Juniperus chinensis* | Pinaceae | |
| Justicia | *Justicia gendarussa* | Acanthaceae | |
| Crape myrtle | *Lagerstroemia indica* | Lythraceae | |
| Lantana | *Lantana camara* | Verbenaceae | |
| Mehndi (Henna) | *Lawsonia inermis* | Lythraceae | |
| Honey suckle | *Lonicera fragrantissima* | Caprifoliaceae | |
| Himchampa | *Magnolia grandiflora* | Magnoliaceae | |
| Barbadose cherry | *Malpighia glabra* | Malpighiaceae | |
| Tapioca | *Manihot utilissima var. variegata* | Euphorbiaceae | |
| Kamini | *Murraya exotica* | Rutaceae | |
| Dhoby bush | *Mussaenda corymbosa* | Rubiaceae | |

| *Common Name* | *Botanical Name* | *Family* |
|---|---|---|
| Vilayati mehandi | *Myrtus communis* | Myrtaceae |
| Kaner | *Nerium indicum* | Apocynaceae |
| Harisingar (Night Jasmine) | *Nyctanthes arbor-tristis* | Oleaceae |
| Snow bush | *Phyllanthus nivosus* | Euphorbiaceae |
| Lalchitra | *Plumbago rosea* | Plumbaginaceae |
| Pomegranate | *Punica granatum* | Punicaceae |
| Corel plant | *Russelia juncea* | Scrophulariaceae |
| Chandi | *Tabernaemontana coronaria* | Apocynaceae |
| Cape honey suckle | *Tecomaria capensis* | Bignoniaceae |
| Pilakaner (Yellow Oleander) | *Thevetia nerifolia* | Apocynaceae |
| Sage rose | *Turnera ulmifolia* | Turneraceae |
| Monk's pepper tree | *Vitex agnus-castus* | Verbenaceae |
| High-bush | *Vaccinium corymbosum* | Ericaceae |

# Appendix X

## A. List of Important Garden Trees

| *Common Name* | *Botanical Name* | *Family* |
|---|---|---|
| Gulabi kachnar | *Bauhinia purpurea* | Leguminosae |
| Samundar-ka-phul | *Barringtonia racemosa* | Lecythidaceae |
| Kachnar (Baisakhi) | *Bauhinia variegata* | Leguminosae |
| Simul | *Bombax malabaricum* | Bombacaceae |
| Dhak or palas | *Butea monosperma* | Leguminosae |
| Camellia | *Camellia japonica* | Theaceae |
| Amaltas | *Cassia fistula* | Leguminosae |
| Pink shower | *Cassia grandis* | Leguminosae |
| Bhockar (Searlet Cordia) | *Cordia sebestena* | Boraginaceae |
| Tope gola | *Couroupita guianensis* | Lecythidaceae |
| Bilasi (Caper tree) | *Crataeva religiosa* | Capparidaceae |
| Rudraksh | *Elaeocaypus ganitrus* | Elaeocarpaceae |
| Pangri | *Erythrina indica* | Leguminosae |
| Easter tree | *Holarrhena antidysenterica* | Apocynaceae |
| Torch tree | *Ixora parviflora* | Rubiaceae |
| Pride of India | *Lagerstroemia speciosa* | Lythraceae |

| Common Name | Botanical Name | Family |
|---|---|---|
| Indian lilac | *Lagerstroemia indica* | Lythraceae |
| Himchampa | *Magnolia grandiflora* | Magnoliaceae |
| Cajeput | *Melaleuca leucadendron* | Myrtaceae |
| Nagkesar (Iron wood tree) | *Mesua ferrea* | Guttiferae |
| Champa | *Michelia champaca* | Magnoliaceae |
| Akasneem | *Millingtonia hortensis* | Bignoniaceae |
| Vilaiti kikkar | *Parkinosia aculeate* | Leguminosae |
| Chameli (Red Jasmine) | *Plumeria rubra* | Apocynaceae |
| Gulmohar | *Poinciana regia (Delonix region)* | Leguminosae |
| Kanak-champa | *Pterospermum acerifolium* | Sterculiaceae |
| Ashok | *Saraca indica* | Leguminosae |
| Agasti | *Sesbania grandiflora/ Agastii grandlora* | Leguminosae |

## B. Foliage (Shade) and Ornamental Trees

| Common Name | Botanical Name | Family |
|---|---|---|
| Monkey-bread tree | *Adansonia digitata* | Bombacaceae |
| Siris tree | *Albizzia lebbek* | Leguminosae |
| Kadamba | *Anthocephalus cadamba* | Rubiaceae |
| Neem | *Azadirachta indica* | Meliaceae |
| Umbrella tree | *Brassaia actinophylla* | Araliaceae |
| Sultan champa | *Calophyllum inophyllum* | Guttiferaceae |
| Beet wood (She-Oak) | *Casuarina equisetifolia* | Casuarinaceae |
| Shisam (Sisoo) | *Dalbergia sissoo* | Leguminosae |
| Chalta | *Dillenia indica* *Eucalyptus multiflora* | Dilleniaceae |
| Bar (Banyan) | *Ficus benghalensis* | Moraceae |
| Java fig | *Ficus benjamina* | Moraceae |
| Indian Rubber Plant | *Ficus elastic* | Moraceae |
| Pakur | *Ficus infectoria* | Moraceae |
| Pipal | *Ficus religiosa* | Moraceae |

| Common Name | Botanical Name | Family |
|---|---|---|
| Bilaiti imli Jalebi | *Inga dulcis*<br>*Juniperus chinensis* | Leguminosae |
| Jhar phanoos (Sausage Tree) | *Kigellia pinnata* | Bignoniaceae |
| Mohua | *Madhuca indica* | Sapotaceae |
| Eucalyptus | *Eucalyptus multiflora* | Myrtaceae |
| Juniperus | *Juniperus chinensis* | Pinaceae |
| Mulsari | *Mimusops elengi* | Sapotaceae |
| Chir (Indian Pine) | *Pinus longifolia* | Pinaceae |
| Ashoka | *Polyalthia longifolia* | Anonaceae |
| Karanj | *Pongamia glabra* | Leguminosae |
| Wild Olive | *Putranjiva roxburghii* | Euphorbiaceae |
| Arjun | *Terminalia arjuna* | Combretaceae |
| Indian Tropical almond | *Terminalia catappa* | Combretaceae |
| Morponkhi | *Thuja occidentalis* | Pinaceae |

# Glossary

**Abaxial:** The side of an organ turned away from the axis, for example, the under side of a leaf.

**Aberrant:** Different from the normal.

**Abnormal:** Different from the usual or prevailing form or structure.

**Abortion:** Failure of development or imperfect development.

**Abortive:** Imperfectly developed or underdeveloped.

**Abruptly pinnate:** Pinnate but without terminal leaflet.

**Abscission:** The shedding of some part or structure as a leaf or flower at its base that break apart readily.

**Acaulescent:** Apparently without a stem that is with the main stem underground and only basal leaves and slender, leafless flowering stems appearing above ground level.

**Accessory:** Additional beyond the normal; extra.

**Accrescent:** Increasing in size after flowering, enlarging with age.

**Accumbent:** Applied to cotyledons with their edges turned towards the main axis of the embryo.

**Aceriform :** Similar to the leaf of a maple.

**Acerose:** Needlelike.

**Achene:** A dry one seeded fruit with a firm close fitting wall which does not open by any regular dehiscence.

**Achene beak:** The persistent and hardened style of an achene.

**Achlamydeous:** Having neither calyx or corolla.

**Acicular:** Needle shaped and very slender.

**Acorn:** The leathery fruit of an oak, containing a single large seed and enclosed basally in a cup formed from bracts.

**Actinomorphic:** Radially symmetrical, that is, capable of division by straight boundaries into three or more similar sections.

**Active bud**: A growing bud.

**Aculeate:** Covered with prickles.

**Acuminate:** With a long, tapering point set off rather abruptly from the main body.

**Acute:** With a pointed end forming an acute angle, that is, less than a right angle.

**Adaxial:** On the side towards the axis, as for example,the upper side of a leaf.

**Adherent:** Grown fast to a structure of a different bind, as for example, a stamen grown fast to a petal.

**Adnate:** Adjective of Adnation.

**Adnation :** Fusion with or attachment to another structure from the beginning of development.

**Adventitious:** A structure occurring in an unusual position as, for example, an adventitious root formed from a stem or a leaf.

**Adventitious bud:** A bud produced in a unusual or unexpected place, as for example near the point of injury of a stem or on leaf or a root.

**Adventitious roots:** Are those shoots that arise from the aerial plant parts, underground stem or from old roots.

**Adventitious shoots:** Are those shoots that develop on roots or inter-nodally on stems.

**Adventive:** Said of an introduced plant beginning to spread into a new locality or region.

**Aerial root:** Roots exposed to the air, as those of tropical plants growing on the trees.

**Aestival:** Of the summer.

**Aestivation:** Arrangement of the young flower parts in the bud.

**Agglomerate:** Crowded into a dense cluster but not joined synonymus aggregate.

**Aitinomic:** As referred to parthenocarpy, the ability of a part to develop parthenocarpic fruits only in response to some stimulus external to ovary.

**Alate :** With a wing.

**Albumen:** A deposit of reserve food material accompanying the embryo. Such reserves often are in the endosperm.

**Alga:** A seaweed or one of the essentially microscopic pond scums. Plural algae.

**Alliaceous:** Having the odor or taste of garlic or onions.

**Alternate:** With a single structure of each bind occurring at each level of the axis appearing to alternate on opposite sides of the axis, but actually in a spiral arrangement.

**Alveolate:** With angular depressions forming a pattern like honey comb.

**Alveolus:** A depression with in an alveolate pattern.

**Ament:** A soft, usually scaly spike of small, apetalous, unisexual flowers, this usually falling as a single unit.

**Amentiferous:** With the flowers in aments *i.e.*, catkins.

**Amorphous:** With no definite form.

**Amphitropous:** Describing an ovule bent back along and adnate to the funiculus, but with the micropyle not bent all the way back to the funiculus.

**Amplexicaul:** Clasping the stem, that is, the base nearly surrounding it.

**Ampliate:** Enlarged.

**Anastomosing:** Joining each other and forming a network, the term being applied to veins as the veins of the leaf.

**Anatropous:** Descriptive of an ovule in which the body is bent backward along the funiculus and adnate to it.

**Ancipital:** Two edged.

**Androecium:** The male reproductive organs; in the flowering plants the stamens- a collective term for all those within a flower.

**Androgynous:** A term applied to an inflorescence which includes both staminate and pistillate flowers, the staminate ones being apical, the pistillate basal.

**Androphore:** A stalk or other supporting structure under stamens raised above their normal position in the flower.

**Anemorphilous:** With the pollen carried by wind.

**Annual:** A plant completing its life cycle in a year or less.

**Annular:** Ring like.

**Annular thickening:** Thickening of the wall of a xylem cell laid down in the form of a ring.

**Annulus:** A ring. In the ferns a crest like or other special structure on the sporangium.

**Anther:** The portion of the stamen which produces pollen.

**Anther tube:** A tube formed by coalescent anthers, as in the sunflower family.

**Antheridium:** A structure producing antherozoids, that is male gametes or sex cells.

**Antherozoid:** A male gamete (sex) cell of a plant.

**Anthesis:** The time when the flower expands and opens or the process of expansion and opening.

**Anthocyanin:** Any of a common class of pigments having colours ranging from labender to purple. These pigments are affected by the acidity or alkalinity of the cell sap, and they change colour in approximately the same way as litmus paper, tending towards red in an acid medium and blue in a basic medium. These pigments are water soluble and not associated with plastids.

**Antipodal cells:** Three cells at the end of the flowering plant megagametophyte opposite the egg.

**Antrorse:** Directed upward or forward.

**Apetalous:** Without petals.

**Aphyllous:** Without leaves.

**Apical:** Of the apex.

**Apical placenta:** A placenta at the distal (apical) end of the ovary.

**Apiculate:** Terminated by an abrupt, short, flexible point.

**Apiculation:** An abrupt, short, flexible point.

**Apocarpous:** With the carpels separate.

**Apogamous:** Developed without the joining of gametes, that is asexually.

**Apomixes:** In some species the embryo is not produced as a result of meiosis and fertilization, but form a cell in the embryo sac or surrounding nucellus which does not undergo meiosis but develops to form a zygote of same genetic makeup as the female parent.

**Apophysis:** A swelling on the surface of an organ as, for example, the somewhat swollen exposed portion of the cone scale of a gymnosperm.

**Appendiculate:** With a basal appendage.

**Appressed:** Lying tightly against another (Usually larger) organ.

**Apterous:** Without wings.

**Aquatic:** Growing in water.

**Arachnoid:** With slender, tangled hairs resembling the threads of a spider web.

**Arboreous:** Of tree like form, that is with a main woody trunk.

**Arborescent:** Of large size and more or less tree like but without the clear distinction of a single trunk.

**Archegonium:** A flask shaped organ containing an egg, that is, female gamete or sex cell.

**Arctic:** The area beyond timber line at high latitudes.

**Arcuate:** Forming a moderate curve or arc.

**Areolate:** Divided into small, marked-off spaces (Reticulate).

**Areole:** Diminutive of area; a small, clearly marked space. The term is used most frequently in reference to the small, special spine-bearing areas on the stem of a cactus.

**Aril:** A large appendage of the funiculus at the hilum (attachment area) of a seed. It tends to envelope the seed.

**Arista:** A stiff bristle.

**Aristate:** With a stiff bristle.

**Articulate:** With conspicuous segments or joints.

**Artificial system:** A system made up according to predetermined supposedly important characters.

**Ascending:** Arising at an oblique angle.

**Aseptic:** Absence of fungi, bacteria, viruses, mycoplasmas, or other microorganisms in cultures.

**Asperous:** Rough.

**Assurgent:** Ascending.

**Altenuate:** With a long, tapering point, this usually set off rather abruptly from the main body of the object (ex. a leaf blade).

**Auricle:** An appendage shaped like the lobe of a human ear.

**Auriculate:** Having an artricle.

**Austral:** Southern.

**Autogamy:** When a flower is fertilized by its own pollen.

**Autonomic:** As referred to parthenocarpy, the ability to set fruits without the stimulus resulting from pollination.

**Awl-shaped:** With a narrow flattened body tapering very gradually upward into a point (subulate).

**Awn:** A long stout or stiff bristle.

**Awned:** With an awn.

**Axenic:** Totally free from association with other organisms.

**Axil:** The adaxial angle between two organs, particularly the angle between the upper side of a leaf and the stem.

**Axile:** On the axis, as for example, said of placentae at or near the centre of an ovary.

**Axillary:** In the axil.

**Axillary bud:** A bud in a leaf axil.

**Axillary flower:** A flower in the axil of a leaf or a bract.

**Auxin:** Auxin is a generic term for compounds characterized by their capacity to induce elongation in shoot cells. They resemble indole-3-acetic acid in physiological action. Auxin may, and generally do, affect other processes besides elongation but elongation is considered critical. Auxins are generally acids with an unsaturated cyclic nucleus or their derivatives.

**Baccate:** Berry like, that is, flushy or pulply.

**Banner:** The upper and usually largest petal in the papilionaceous corolla of a plant of the pea family. Known also as a standard or a vexillum.

**Barbed:** With a rigid barb like the barb of a fish hook.

**Barbellate:** Diminutive of barbed (Still smaller, barbellulate).

**Barbulate:** With a beared of fine hairs.

**Bars of sanio:** Special thickening on the walls of some xylem cells.

**Basal leaves:** Leaves at the base of an herbaceous plant arising from several nodes separated by exceedingly short internodes occurring at about ground level.

**Basal placenta:** A placenta at the basal end of the ovary.

**Basifixed:** Attached at the base.

**Beaked:** Ending in a firm elongated slender structure.

**Beared:** With long or stiff hairs.

**Berry:** A fleshy or pulpy fruit with more than one seed and formed from either a superior or an inferior ovary. The seeds are embedded in pulpy tissue.

**Bidentate:** With two teeth.

**Biennial:** Completing the life cycle in two years. Biennial plants usually produce only basal leaves above the ground the first year and both basal leaves and flowering stems the second.

**Bifid:** Forked, that is, ending in two parts.

**Bilabiate:** Two lipped. A bilabiate corolla has petals in two sets commonly with two in the upper and three in the lower.

**Bilateral:** Two sided.

**Bilaterally symmetrical:** Capable of division into only two similar sections, which are mirror images of each other.

**Bilocular:** With two cavities, as an ovary with two seed chambers.

**Bipinnate:** Pinnate and the primary leaflets again pinnate.

**Bipinnatified:** Pinnatified with the primary divisions again pinnatified.

**Bisexual:** With both sexes represented in the same individual or organ *e.g.*, with stamens and pistils in the same flower.

**Blade:** The broad, usually flat part of a leaf.

**Bloom:** A usually waxy; whitish or bluish powder covering the surface of a leaf, stem, fruit or other organ.

**Blossom end:** The end of an inferior ovary and the surrounding floral cup which supports the flower parts or at fruiting time the calyx and remains of other flower parts.

**Boreal:** Northern.

**Boss:** A protrusion, as for example, a protuberance near the centre of the apophysis of a scale of a gymnosperm cone.

**Bract:** A leaf subtending a reproductive structure, such as a flower or a cluster of flowers or an ovuliferous scale. Usually the leaf is specialized and at least somewhat dissimilar to the foliage leaves.

**Bracteate:** With bracts.

**Bracteolate:** With bractlets.

**Bracteole:** Diminutive of bract, a mere scale.

**Bractlet:** A minor bract.

**Branchlet:** A small branch, that is one of the smallest degree.

**Branch primordium:** The lump of tissue which develops into a young branch.

**Bristle:** A stiff hair.

**Bryophyta:** The mosses, liverworts and anthocerotae.

**Bud:** A growing structure at the tip of the stem or a branch, with the enclosing scale leaves or immature leaves; a young flower which has not yet opened. There are vegetative buds and flower buds, as described in the definition above.

**Bulb:** An underground bud covered by fleshy scales, the coating formed from the bases of leaves.

**Bulbiferous:** With bulbs.

**Bulbils:** A small bulb. The word is applied to bulb like structures produced on the stems, usually in the axils of leaves or sometimes in the places where flowers ordinarily occur.

**Bulblets:** A small underground bulb.

**Bulbous:** Bulblike.

**Bullate:** Blistered or puckered.

**Caducous:** Falling very early. In the poppy family the caduceus sepals fall away when the flower open.

**Caespitose:** Growing in tufts or mats.

**Calcarate:** With a spur.

**Callosity:** A thickened hard structure.

**Callous:** With the texture of a callus.

**Callus:** Production of new unspecializing cell from a part of the plant, growing as a result of injury or wounding.

**Calyculate:** With small bracts in a series below the flower, these resembling a calyx.

**Calyptra:** A cap, cover or lid.

**Calyx:** A cup; the sepals of a flower, the outermost series of flower parts. Each sepal has the same number of vascular traces as a leaf of the species. If there is doubt it may be distinguished by this character cf petal.

**Calyx tube:** A tube formed from the lower portions of the sepals. The term has been used loosely for (a) a floral cup or floral tube regardless of its origin, (b) a floral cup or tube formed by coalescence and adnation of the bases of the sepals, petals and stamens, (c) a perianth tube of the type in the monocotyledons formed by edge to edge adnation of the adjacent sepals and petals together with the bases of the stamens (d) and the portion of a floral cup or tube above the area of adnation to an inferior ovary.

**Cambium:** A group of cells which possess the power of active growth and division and which is responsible for the production of old tissues in matured organs.

**Campanulate:** Bell shaped, that is, the shape of an inverted church bell *i.e.,* rounded at the attachment and with a broad flaring rim.

**Campylotropous:** Descriptive of an ovule which curves in such a way that the micropyle is near the funiculus or stalk but the side of the ovule is not adnate to the funiculus.

**Canaliculated:** With lengthwise furrows or channels.

**Cancellate:** Appearing to be a lattice work.

**Canescent:** Grayish- white or hoary, densely covered with white or gray fine hairs, these usually short.

**Capillary:** Like an elongated delicate hair or thread.

**Capitate:** In a dense cluster or head.

**Capitulate:** Diminutive of capitate.

**Capitulum:** A small head.

**Caprifig:** The wild or 'male' fig, the uncultivated form.

**Capsule:** A dry, many seeded fruit made up of more than one carpel and splitting open length wise at maturity.

**Carina:** A keel, literally like the keel of a boat. Sometimes said of a structure which simply is folded and creased and without a projecting keel on the crease.

**Carinate:** With a keel.

**Carpel:** A specialized leaf which forms either all or part of a pistil.

**Cartilaginous:** With the texture of cartilage that is, tough and firm but somewhat flexible.

**Caruncle:** An appendage at the attachment point of a seed.

**Carunculate:** With a caruncle.

**Caryopsis:** The fruit of a grass that is, a one seeded, indehiscent fruit with the pericarp adnate to the seed.

**Castaneous:** Chestnut coloured.

**Catkin:** A soft, usually scaly spike or raceme of small apetalous unisexual flowers, the inflorescence usually falling as a single unit.

**Caudate:** With a talelike structures.

**Caudex:** A largely underground stem base which persists from year to year and each season produces leaves and flowering stem of short duration.

**Caulescent:** Having a well developed stem above ground as opposed to having only what appears to be the stalk of the individual flower or cluster of flowers.

**Cauline:** Of or on the stem.

**Cell:** The structural and functional unit of living organisms.

**Cell culture:** This term is used to denote the growing of cells *in vitro* including the culture of single cells. In cell cultures, the cell are no longer organized into tissues.

**Cell hybridization:** The fusion of two or more dissimilar cells leading to the formation of a synkaryon.

**Cell line:** A cell line arises from a primary culture at the time of the first successful subcultures. The term cell line implies that cultures from it consist of numerous lineages of cells originally present in the primary cultures. The terms finite or continuous are used as prefixes if the status of the culture is known. If not, the term line will suffice. The term "continuous line" replaces the term "established line". In any published description of a culture, one must make every attempt to publish the characterization of the history of the culture. If such has already been published, a reference to the original publication must be made. In obtaining a culture from another laboratory, the proper designation of the culture, as originally named and described, must be reported in any publication.

**Cell strain:** A cell strain is derived either from a primary culture or a cell line by the selection or cloning of cells having specific properties or markers. These properties or markers must persist during subsequent cultivation. In describing a cell strain, its specific features must be defined. The terms finite or continuous are used as prefixes if the status of the culture is known. If not, the term strain will suffice. The term "continuous line" replaces the term "established line". In any published description of a cell strain, one must make every attempt to publish the characterization of the history of the strain. If such has already been published, a reference to the original publication must be made. In obtaining a culture from another laboratory, the proper designation of the culture, as originally named and described must be maintained and any deviations in cultivation from the original should be reported in any publication.

**Centrifugal:** Developing first at the centre and then gradually toward the outside.

**Centripetal:** Developing first at the outside and the gradually towards the centre.

**Cernuous:** Nodding.

**Certified seed:** Progeny of the registered seed that is produced in the largest volume and sold to the crop producers.

**Chaff:** Dry, membranous scales or bracts, that is, similar to the chaff which comes from a thrashing machine.

**Chaffy:** Resembling chaff.

**Chamber:** A room. Applied to the cavities of an anther or an ovary.

**Channeled:** With lengthwise grooves, the grooves usually deep.

**Chartaceous:** Like writing paper.

**Chasmogamy:** Pollination only after the opening of the flower.

**Chimeras:** When a mutation occurs within a single cell of a clone it initially produces an "islands" of mutant cells within a growing point of a stem. If successful, the plant then becomes a mixture of two different genotypes. This structural arrangement is known as a chimera.

**Chlorophyll:** The green colouring of most plants. A substance which aids in making the energy of light available to photosynthesis.

**Chloroplast:** Literally a green body, actually a solid body with chlorophyll in its outer part.

**Chlorosis:** A diseased condition shown by loss of green colour.

**Choripetalous:** With the petals separate or atleast with some of them separate from others.

**Cilia:** Hairs along the margin of a structure, placed like the eye-lashes on a human eyelid.

**Ciliate:** With cilia along the margin.

**Ciliolate:** Diminutive of ciliate.

**Cinereous:** The colour of ashes.

**Circinate:** Coiled at the tip, as the young coiled leaf of a fern which uncoils gradually as it develops.

**Circumscissile:** Opening by a horizontal circular line, the top coming off like a lid.

**Class:** The next taxon below division; a group of related orders or sometimes a single order.

**Clavate:** Gradually enlarged upward after the manner of a base-ball bat or the traditional giants club, that is, either tapering gradually upward or with an enlarged knob at the summit.

**Clavellate:** Diminutive of clavate.

**Claw:** The narrow stalk at the base of a petal. This resembling the petiole of a leaf.

**Cleft:** Indented about half way or little more than half way to the base or to the midrib.

**Cleistogamy:** Self-pollination without the flower opening.

**Cleistogamous:** Self-fertilized in the bud stage or atleast without opening.

**Climbing:** Supported by clinging (Scandent).

**Clonal propagation:** A sexual reproduction of plants that are considered to be genetically uniform and originated from a single individual or explant.

**Clone:** Genetically uniform material derived from a single individual and propagated exclusively by vegetative means as cuttings, graftage or division.

**Coalescence:** The union of similar parts as for example, the petals of a flower.

**Coalescent:** Adjective.

**Cochleate:** Spiral, like a turban snail.

**Coherent:** Grown fast together.

**Colonial:** In colonies. For the angiosperms used, primarily in reference to plants occurring in clumps connected by rhizomes.

**Coloure:** The failure of blossoms to set, resulting in a pre-matured drop.

**Column:** A slender aggregation of coalescent stamen. Filaments, as in some members of the mallow family.

**Coma:** A tuft of hairs on the end of a seed.

**Commissure:** The surface along which two or more locules are joined to each other.

**Comose:** With a coma.

**Compatible:** Ability of pollen to develop in the styles rapidly enough to reach the ovule in time to effect fertilization.

**Compatibility:** Of sex cells, the ability to unite and form a fertilized egg that can grow to maturity.

**Complete flower:** A flower with all the four usual series that is sepals, petals stamen and pistils.

**Compound:** Composed of two or more similar elements.

**Compound leaf:** It is composed of two or more leaflets, a compound pistil is composed of two or more coalescent carpels.

**Compressed:** Flattened, particularly from side to side.

**Conduplicate:** Folded together lengthwise.

**Cone:** A reproductive structure composed of an axis bearing sporophylls or other seed or pollen bearing structures.

**Congeniality:** As determined by the degree of success of the union between stock and scion.

**Conglomerate:** Densely aggregated; a cluster or a heap.

**Coniferous:** Cone bearing.

**Conjugate:** Lying together in pairs.

**Connate:** Joined from the beginning of development.

**Connate perfoliate:** Both connate and perfoliate.

**Connective:** The middle part of an anther connecting the two pollen sacs or two pairs of pollen sacs.

**Connivent:** Standing together. Stamens with their tips ending against each other are connivent.

**Contorted:** Twisted out of the usual form. Petals may be contorted in the bud.

**Convolute:** Rolled up lengthwise. Petals may be convolute in the bud.

**Cordate:** Of a conventional heart-shape, the length greater than the width, the petiole attached in the basal sinus; applied also to the basal indentation of a leaf or other structures.

**Coriaceous:** Leathery.

**Corm:** A bulblike structure formed by enargement of the stem base. It is sometimes coated with one or more membranous layers.

**Corneous:** Horny.

**Corolla:** The petals of a flower, that is, the inner series (or several series) of the parianth. Each petal has usually a single vascular trace as does a stamen.

**Corolla division:** A division is longer than a lobe or part.

**Corolla lobe:** A lobe is longer then the tooth. Lobes may be applied in a broad sense to cover lobes, parts or divisions.

**Corolla part:** A part is longer than a lobe.

**Corolla teeth:** The separate tips of the individual petals of a sympetalous corolla.

**Corolla tube:** The hollow cylinder formed by coalescence of petals.

**Corona:** A small crown, the term being applied to special appendages of the corolla. These sometimes form a tube.

**Coroniform:** Shaped like a crown.

**Corrugated:** With many small folds or wrinkles.

**Cortex:** The layers of living cells outside the central stele but inside the outermost single layer of cells of a stem or a root.

**Cortical:** Of the cortex.

**Corymb:** A flat topped cluster of flowers; fundamentally like a raceme but with the pedicels of the lower flowers longer and the pedicels of the upper flowers gradually shorter.

**Corymbiform:** In the form of a corymb.

**Corymbose:** Arranged in corymbs.

**Costa:** A rib or a promiment nerve.

**Costate:** Ribbed lengthwise.

**Cotyledons:** One of the first leaves developed in the embryo in the seed at the joining point of the hypocotyls and the epicotyl. Often food materials are stored in them.

**Creeping:** The stem growing along the ground and producing adventitious roots.

**Crenate:** With rounded teeth projecting at right angles to the edge of the leaves.

**Crenulate:** Diminutive of crenate.

**Crested:** With a crest, that is a projection at a prominent position such as the apex or the midrib of an organ.

**Crinkle:** A disorder of apples in which the surface of the fruit becomes roughened. Supposed to be a form of draught injury.

**Crisped:** Ruffled, that is with the same bind of winding as in sinuate but in the vertical plane instead of the horizontal.

**Cristate:** Crested.

**Cristulate:** The diminutive of cristate.

**Cross-pollination:** Transfer of pollen grain from a flower of one variety to a flower of another variety.

**Cross-compatible:** The pollen of one variety 'A' is capable of functioning in the styles and fertilizing the ovules of variety 'B' (variety 'B', however, may not be cross-compatible with 'A').

**Cross-incompatible:** Variety 'A' produces functional sex cells, but its pollen tube grows too slowly in the styles of variety 'B' to effect fertilization. Variety 'A' may, however, serve as an effective pollinizer for some other varieties. (Variety 'B' may serve as an effective pollinizer for variety 'A').

**Cross-fertile:** One variety 'A' is used as a pollinizer for another variety 'B' and 'B' produces fruit with viable seed.

**Cross-fruitful:** One variety 'A' is used as a pollinizer for another variety 'B' and 'B' produces a commercial crop.

**Cross-pollination:** The transfer of pollen from the anthers of a flower of one variety to the stigma of a flower of a different variety.

**Cross-unfruitful:** One variety 'A' is used as a pollinizer for another variety 'B', and 'B' fails to produce a commercial crop.

**Cross-sterile:** One variety 'A' is used as a pollinizer for another variety 'B' and 'B' fails to produce fruit with viable seed.

**Cross section:** A slice cut across an object.

**Cruciate, cruciform:** Cross like.

**Crustaceous:** Hard and brittle in texture.

**Cryptogames:** An old term applied to the thallophytes, bryophytes and the pteridophytes, that is the plants which regularly reproduce with out formation of seeds.

**Cucullate:** Hood shaped

**Culum:** The hallow stem of a grass, the term being applied some times to ridges also.

**Cuneate, cuneiform:** Wedge shaped; essentially a narrow isosceles triangle with the distal corners rounded off. The petiole of a cuneate leaf is attached at the sharp angle.

**Cupule:** A little cup.

**Cuspidate:** With a sharp, firm point at the tip, as applied to leaves, the term indicates a point of firmer texture than the rest of the blade.

**Cuticle:** The waxy, more or less waterproof coating secreted by the cells of the epidermis of a leaf, stem or flower parts.

**Cyathiform:** Cup shaped.

**Cyathium:** A cuplike involucre enclosing flowers.

**Cycle:** A circle. Leaves or flower parts arranged in a single series at a single node or in a cycle.

**Cyclic:** Arranged in a cycle.

**Cylindroidal:** In approximately the shape of a cylinder.

**Cyme:** A broad, more or less flat-topped cluster of flowers.

**Cymose:** With cyme like cluster of flowers.

**Cymule:** Diminutive of cyme.

**Cypsela:** An achene which is attached to the enclosing floral cup or tube, that is, an achene formed from an inferior ovary.

**Cytology:** The study of cell. Often particular emphasis is placed upon the chromosomes.

**Cytoplasm:** The living substance outside the nucleus and inside the enclosing membrane.

**Deciduous:** Falling off at the end of each growing season.

**Deciduous tree:** Tree which shed their leaves during winter months to enable them to stand cold injury.

**Declinate:** Bending or curving downward or forward.

**Declined:** Bending over in one direction.

**Decompound:** More than one compound.

**Decombent:** Reclining except at the apex.

**Decurrent:** Leaf bases which continue along the stem as wings or lines.

**Decussate:** In opposite pairs with the alternate pairs projecting at right angles to each other.

**Deflexed:** Abruptly bent or turned downward.

**Defined medium:** A nutritive solution for culturing cells in which each component is of known chemical structure. Although it is recognized that even the "Purest" chemical compounds may have some contaminants, high quality chemicals should be used with analytic data, if possible, on contaminants.

**Dehisce:** The split open along definite lines.

**Dehiscence:** The process of splitting open at maturity.

**Dehiscent:** Splitting open along definite lines.

**Deltoid:** Of the shape of the greek letter delta, that is, an equilateral triangle, the attachment being in the middle of one side.

**Dentate:** With angular teeth projecting at right angles to the edge of the structure.

**Denticulate:** Diminutive of dentate.

**Depauperate:** Stunted, that is, very small.

**Depressed:** Flattened from above as if pushed downward.

**Determinate:** With a definite predetermined number of structures.

**Diadelphous:** In two brotherhoods, the term being applied to stamens coalescent in two sets.

**Diandrous:** With two stamens.

**Dichasium:** A cyme with two axis running in opposite directions, that is, the type of cyme formed in plants with opposite branching in the inflorescence. See cyme.

**Dichlamydeous:** With both calyx and corolla.

**Dichogamy:** Insuring cross-pollination by the sexes being developed at different times.

**Dichotomous:** Forking, with two usually equal branches at each point of forking.

**Dicotyledonous:** With two cotyledons.

**Dicliny:** Male and female organ separate and in different flowers.

**Didymous:** Twin like, that is, occurring in pairs.

**Didynamous:** In two pairs, the pairs not being of the same length. The term is applied to stamen.

**Diffuse:** Spreading widely and diffusely in all directions.

**Digitate:** Resembling the fingers of a human hand, that is, with several similar structures arising at a common point. See palmate.

**Digynous:** With two pistils.

**Dimerous:** With two members; the flower parts in two.

**Dimidiate:** With the appearance of being only half a structure.

**Dimorphous:** With two forms.

**Dioecious:** Unisexual, the male and female elements in different individuals.

**Diploid:** With 2n chromosomes per cell.

**Diplostemonous:** The stamens in two series, those of outer series alternating with the petals.

**Dipterous:** With two wings.

**Disc, disk:** In the compositae, the central portion of the compound receptacle bearing the disc flowers.

**Disc flowers:** One of the flowers with tubular corollas.

**Disciform:** Circular and flattened like disc.

**Discoid:** Resembling a disc.

**Discrete:** Separate.

**Dissected:** Divided into narrow segments.

**Dissepiment:** A portion dividing an ovary or a fruit into chambers.

**Distichous:** In two vertical series.

**Distinct:** Separate.

**Diurnal:** Occuring in the day time. The term is applied to the flowers which open in day light as opposed to the nocturnal flowers which open at night.

**Divaricate:** Spreading widely, that is, divergent.

**Divergent:** Spreading away from each other.

**Divided:** Indented essentially to the base or midrib.

**Division:** The highest rank of taxon in the plant kingdom, a group of related classes or some times a single class; a segment of a structure, such as leaf.

**Dormant:** Applied to buds when they are not actively growing and to plants when they are not in leaf.

**Dorsal:** On the outer surface of an organ, that is, the side away from the axis.

**Dorsiventral:** A structure having a clear differentiation of a back and front or upper and lower side.

**Dorsoventral:** The distance or measurement from the back of a structure to the front, that is, from dorsal to ventral, as opposed to lateral distance.

**Double samara:** A Samara with two locules and two wings.

**Downy:** Finely and softly pubescent.

**Drainage area:** The watershed or area into which the excess precipitation of a region drains.

**Drupaceous:** Druplike.

**Drupe:** A fruit with a fleshy exocarp and a hard, stony endocarp about each seed. The term is used also for other fruits which are fleshy or pulpy on the outside and with one or more stones inside.

**Drupelet:** Diminutive of drupe. The cluster of fruits of a blackberry or raspberry is composed of druplets which seem to form a single fruit.

**Echinate:** Covered with prickles.

**Egg:** A female gamete cell of a plant.

**Effective bloom:** The length of time the tree is in conspicuous blossom.

**Ellipsoid, ellipsoidal:** Elliptical in outline with three-dimensional body.

**Elliptic, elliptical:** In the form of an ellipse, that is, about one and one half times as long as broad, widest at the middle and rounded at both ends. A two dimensional figure.

**Emarginate:** With a shallow broad notch at the apex.

**Embryo:** The new plant enclosed in the seed. Its formation follows union of gametes.

**Embryo culture:** *In vitro* development or maintenance of isolated mature or immature embryos.

**Embryogenesis:** The process of embryo initiation and development.

**Embryo sac:** The cell in the ovule in which the embryo is formed.

**Emersed:** Above water.

**Enation:** An outgrowth from the superficial tisssues of the stem.

**Endimic:** Restricted in occurrence to aparticular geographical area.

**Endocarp:** The inner layer of the wall of a fruitified ovary.

**Endosperm:** A cell layer occurring in at least the immature seeds of flowering plants.

**Ensiform:** Sword like; in the form of sword.

**Entire:** Without division or toothing of any bind.

**Entomophilous:** Insect pollinated.

**Emphemeral:** Lasting for a brief period.

**Epicotyl:** The portion of the embryo of a seed plant just above the cotyledons, the young stem.

**Epigynous:** The sepals, petals and stamens apparently upon the ovary, but actually growing from the edge of the floral cup, which is adnate to the ovary.

**Epigynous disc:** A disc within the floral cup and on top of the inferior ovary.

**Epiphyte:** A plant growing upon another plant but not parasitic upon it.

**Equitant:** With the leaves folded around a stem after the manner of the legs of a rider around a horse or some times with the leaves folded around each other in rows.

**Errect:** Standing upright.

**Erose:** With the margin appearing to have been gnawes.

**Etiolated:** White through failure to develop chlorophyll.

**Euploid:** The situation which exists when the nucleus of a cell contains exact multiples of the haploid number of chromosomes.

**Even pinnate:** Without a terminal leaflet.

**Exocarp:** The outer layer of the wall of a fruitified ovary.

**Excentric:** Off centre.

**Excortis:** A shelling off the bark of tree.

**Excurrent:** Projecting as for example, a leaf base which projects beyond the margin of the blade.

**Exfoliating:** Separating into thin layer.

**Exocarp:** The outer layer of the pericarp.

**Explant:** Tissue taken from its original site and transferred to an artifical medium for growth and maintenance.

**Explanate:** Flattened and spread out.

**Exserted:** Projecting beyond the usual containing structure.

**Extraterritorial:** Occuring beyond the specific geographical range of some features of this book *i.e.*, beyond the north America north of Mexico.

**Extrose:** Facing outward.

**Falcate:** In the shape of a scythe, that is curving, flat and tapering gradually to a point.

**False partition:** An ovary partition not formed by infolding of the edges of the carpel, growing from the middle of the placenta on the opposite side or formed by ingrowth of placentae.

**Farinose:** Mealy, that is composed of mealy granules or with granules on the surface.

**Fasciate:** Literally, in a bundle or bundled together, the term being applied as fascinated, to branches remaining parallel and grown abnormally together.

**Fasciation:** A deformity in which the stem becomes much flattened as a result of multiple terminal buds arranged in a single plane.

**Fascicle:** A bundle or cluster.

**Fastigiate:** Errect and close together.

**Faveolate or favose:** Resembling a honey comb.

**Fecundation:** The fusions of two gametes to form a new cell.

**Fecundity:** The ability of flowers to produce seeds that will germinate.

**Female gamete:** An egg; a female gamete cell.

**Fenestrate:** Perforated.

**Ferruginous:** Rust coloured.

**Fertile:** Productive, mean capable of producing fruit or spores. Sometimes a staminate flower is referred to as infertile or sterile because it produces no seed.

**Fertility:** The ability to set and mature fruit with viable seed.

**Fertilization:** The union of male germ cell, contained in the pollen grain, with the female germ cell, or egg, in the ovule.

**Fibrillose:** With fine fibres.

**Fibrous root system:** A root system with several major roots about equal and arising from approximately the same point.

**Field capacity:** Is defined as the amount of water held against the force of gravity.

**Filament:** A thread; the stalk of a stamen.

**Filamentous:** Composed of threads.

**Filiform:** Thread like, that is, long slender and cylindroidal.

**Fimbriate:** Fringed, that is resembling the fring on the sleeve of an early American buckskin shirt.

**Fimbrillate:** The diminutive of fimbriate.

**Fimbriolate:** With a very fine fringe.

**Fistulose:** Hollow and cylindroidal.

**Flabellate:** Fan shaped.

**Flaccid:** Weak.

**Flagelliform:** Whiplike or lash like.

**Flates:** They are shallow plastic, styrofoam, wooden or metal trays with drainage holes in the bottom. They are useful for germinating seeds or rooting cuttings, since they permit young plants to be moved easily.

**Fleshy fruits:** A fruit with soft, juicy tissues.

**Flexuous:** Curved in first one direction, then the opposite.

**Floccose:** With tufts of wooly hair.

**Flocculent:** Diminutive of floccose.

**Floral cup:** A cup bearing on its rim the sepals, petals and stamens; originating as (a) a hypanthium. (b) a "calyx tube" formed by coalescence and adnation of the bases of the sepals, petals and stamens, or (c) a perianth tube of the type in the monocotyledons, the sepals and adjacent petals being adnate edge to edge.

**Floral tube:** An elongated, slender floral cup.

**Floret:** A small flower, the flower of a grass and the two immediately enclosing bracts, that is the lemma and palea.

**Floricana:** A canelike stem producing flowers.

**Floriferous:** Bearing flowers.

**Flower:** A complex strobilus formed at the end of a branch, including the receptacle (thalamus or torus) and bearing sepals, stamens, petals and pistils or some of these.

**Flower bud:** A bud which will develop into a flower, actually the young flower.

**Flowering hormones:** Are hormones which initiate the formation of floral primodia, or promote their development.

**Flowering regulators:** Are regulators which affect flowering.

**Foliaceous:** Leaflike.

**Foliage bud:** A bud containing new young leaves.

**Foliage leaf:** The green leaf of the mature plant.

**Foliar:** Relating to a leaf.

**Foliate:** Leaved, an example being 3- foliate or trifoliate.

**Foliolate:** With leaflets.

**Foliose:** Bearing many leaves.

**Follicle:** A dry fruit formed from a single carpel, containing more than one seed and splitting open along the suture. The term is applied also sometimes to similar fruits splitting only along the midrib.

**Follicular:** Of the nature of a follicle.

**Forked:** Dividing into branches which are nearly equal.

**Foundation seeds**: Progeny of breeders seed which is so handled as to maintain the highest standard of genetic identity and purity.

**Foveolate:** Pitted.

**Free:** Separate from other organs (distinct).

**Friability:** A term indicating the tendency for plant cells to separate from one another.

**Frond:** The leaf blade of a fern.

**Frondose:** Leafy, that is with frond like leaves.

**Fruit:** A matured ovary with its enclosed seeds and some times with attached external structures.

**Fruitescent:** Shrubby or becoming so at length.

**Fruitful:** A plant or variety that sets and matures a commercial crop of fruit.

**Fruiticose:** Distinctly woody, living over from year to year, and attaining considerable size, with several main stems instead of a single trunk.

**Fruit setting:** A development of the ovary and adjacent tissues following the blossoming period.

**Fugacious:** Falling away early.

**Fulbous:** Tawny.

**Fungus:** A plant without chlorophyll and with no vascular tissue, stems or leaves. Plural fungi.

**Funiculus:** The stalk of an ovule or a seed.

**Funnelform:** In the shape of a funnel.

**Furcate:** Forked.

**Fusicous:** Grayish-brown.

**Fusiform:** In the shape of a spindle, that is, widest at the middle and tapering gradually to each pointed end, the body being circular in cross section.

**Galea:** A hood formed from a portion of the parianth derived, for example, from the two upper petals coalascent indistinguishably into one or from the upper sepal.

**Galeate:** With a galea.

**Gamete:** A unisexual cell which must fuse with another gamete to produce a new individual.

**Gamopetalous:** With all the petals coalescent (Sympetalus).

**Gametophyte:** The gamete producing generation, each cell with a chromosomes.

**Geminiate:** Equal or arranged in pairs.

**Generative cell:** A cell in the microgametophyte (or pollen grain) which may give rise directly or indirectly to male gamete cells or nuclei.

**Genetics:** The study of heredity.

**Geniculate:** Bent like the human knee.

**Genus:** A group of related species or sometimes a single species. Plural, genera.

**Gibbous:** Swollen or distended on one side.

**Glabrate:** At first hairy but latter becoming glabrous.

**Glabrous:** not hairy.

**Gladiate:** Sword like.

**Gland:** A secreting organ.

**Glandular:** Bearing glands, small cellular organs secreting oils, tars, resins etc. often only the secretion is visible.

**Glaucescent:** More or less glaucous.

**Glaucous:** Covered with a white or bluish powder or bloom, this often composed of finely divided particles of wax. Ex. The bloom on a plum.

**Globose, globular:** Spheroidal.

**Glochid:** A sharp hair or bristle tipped with a barb.

**Glochidiate:** Barbed at the tip.

**Glomerate:** In compact clusters.

**Glomerulate:** Diminutive of glumerate.

**Glomerule:** A single cyme of sessile flowers forming a compact cluster. It may be distinguished from a head by blooming of the central flower first.

**Glumaceous:** Resembling a glume of a grass spikelet.

**Glume:** One of the two chaff like bractlets at the base of a grass spikelet. The glumes do not enclose flowers.

**Glutinous:** Covered with sticky material.

**Granulose:** Covered by minute grains of hardened material.

**Gregarious:** In large colonies.

**Growth hormones:** A hormone which regulate growth.

**Growth regulators:** (Synonym: growth sunstances) are regulators which affect growth.

**Gummosis:** A disorder particularly of stone and citrus fruit, in which there are copious exudations or deposites of gum.

**Gum spots:** A disorder of stone, citrus and certain other fruits in which there are small local deposites of gum in the tissues of fruits, shoot or other organ.

**Gymnospermous:** With the seeds naked, that is not enclosed in an ovary.

**Gynandrium:** A structure formed by adnation of stamens and pistil.

**Gynandrous:** With the stamen adhering to the pistils.

**Gynecandrous:** With staminate and pistillate flowers in the same inflorescence, the pistillate above and the staminate below.

**Gynobase:** An enlarged or elongated portion of the receptacle bearing the pistil.

**Gynoecium:** The pistil of a flower or its group of pistils.

**Gynophore:** A special stalk under a pistil.

**Habit:** The general appearance of a plant.

**Habitate:** The type of locality or the set of ecological conditions under which the plant grows.

**Hair:** A slender cellular projection.

**Halberd-shaped:** Hastate.

**Halophyte:** A plant growing in salty soil as for example an alkali flat or a salt water marsh.

**Hamate:** With a hook at the tip.

**Haploid:** With a chromosomes per cell.

**Hastate:** More or less sagitate (arrow head shaped) but with the divergent basal lobes.

**Head:** A cluster of sessile or essentially sessile flowers or fruits at the apex of a peduncle. Essentially a spike with a very short axis.

**Helicoid:** Spiral, like the shell of a turban snail.

**Helicoid cyme:** A coiled inflorescence. The main stem terminates in a flower and a single bud just below grows out as a stem but terminates also in a flower, the process being repeated many times and always in the same direction.

**Herb:** A non woody plant, at least one which is not woody above ground level.

**Herbaceous:** Not woody.

**Herbarium:** A collection of pressed plant specimens.

**Hermaphrodite:** Bisexual, that is, with stamens and pistils in the same flower.

**Heterocarpous:** Producing more than one kind of fruit.

**Heterogamous:** With more than one kind of flower.

**Heterogeneous:** Of various binds, that is, not all individuals the same.

**Heteroploid:** The term given to a cell culture when the cells comprising the culture possess nuclei containing chromosome numbers other than the diploid number. This ia a term used only to describe a culture and is not used to describe the individual cells. Thus, a heteroploid culture would be one which contains aneuploid cells.

**Hilium:** The point of attachment of the funiculus (Stalk) to the seed.

**Hip (of a rose):** A floral cup which usually becomes enlarged and fleshy at fruiting time. The true fruits are achenes inside.

**Hippocrepiform:** Horse shoe shaped.

**Hirsute:** With fairly coarse more or less stiff hairs.

**Hirsutulus:** Slight hirsute.

**Hispid:** With rigid or stiff bristles or bristly hairs.

**Hispidulous:** Diminutive of hispid.

**Hoary:** Covered with short, dense, grayish white hairs, the surface of the stem, leaf or other structure therefore appearing white.

**Homogamous:** With only one kind of flower.

**Homogeneous:** With all members similar.

**Hood:** A hoodlike structure, often formed from a petal or a sepal or more than one of either.

**Humifuse:** Spreading on the ground.

**Hyaline:** Thin and membranous being transparent or translucent.

**Hybrid:** Produced by dissimilar parents.

**Hydrophyte:** An aquatic plant.

**Hygroscopic:** Changing form with changes of moisture content.

**Hygroscopic coefficient:** The percentage of soil water retained in contact with a saturated atmosphere and in the absence of any other source of moisture.

**Hypanthium:** A floral cup or tube developed by extra growth of the margin of the receptacle. An alternative adopted by some authors in use of hypanthium with the meaning of floral cup.

**Hypocotyl:** The portion of the axis of an embryo of a seed plant just below the cotyledon(s).

**Hypogaeous:** Below the ovary, that is, referring to flower parts which come directly from the receptacle and not from a floral cup or tube. A hypogynous flower is one which does not have a floral cup or tube.

**Hypogynous disc:** A flushy cushion of tissue growing from the receptacle below the ovary, often the stamens and sometimes the petals being attached to it.

**Imbricate:** overlapping like the shingles on a roof.

**Imparipinnate:** Odd pinnate.

**Implexed:** Tangled, interlacing.

**Implicate:** Woven in.

**Inactive bud:** A bud dormant either before the beginning of the growing season or through a longer period.

**Incanous:** Covered with white hairs.

**Included:** Not protruding beyond the normal surrounding structures.

**Incompatible:** Inability of viable pollen to develop in the styles rapidly enough to reach the ovule in time to effect fertilization.

**Incompatibility:** Of sex cells, the inability to unite and form a fertilized egg that can be grown to maturity.

**Incomplete flower:** A flower lacking one or more of the four usual series of structures, that is, sepals, petals, stamens or pistils.

**Incrassate:** Thickened.

**Incumbent:** Applied to a pair of cotyledons which lie with the back of one against the axis of the embryo.

**Indehiscent:** Not opening by splitting along regular lines or not opening at all.

**Indeterminate:** Of indefinite growth *i.e.* the size or number of organs to be produced not predetermined.

**Indigenous:** Native in a particular region.

**Indument:** A covering of hairs.

**Induplicate:** With the edges folded inward.

**Indurate:** Hardened.

**Indusium:** The membranous covering of a sorus of a pteridophyte.

**Inferior:** Below.

**Inferior floral cup:** The portion of a floral cup or tube adnate to the wall of an inferior ovary.

**Infertile:** Varieties 'A' and 'B' both produce fruit with viable when pollinated by each other.

**Inflorescence:** The flowering area or segment of a plant.

**Integument:** The outer coating of an ovule, a structure grown up from the sporophyll and surrounding the sporangium. The integument becomes the seedcoat.

**Intercompatible:** The pollen produced by either variety of a combination is capable of functioning in the styles and fertilizing the ovule of the other variety.

**Interfruitful:** Varieties 'A' and 'B' both produce commercial crops when pollinated by each other.

**Interincompatible:** Varieties 'A' and 'B' are unfruitful when pollinated by each other because the pollen tubes of each variety grow too slowly in the styles of the other to effect fertilization. Either variety may serve as an effective pollinizer for some other varieties.

**Intersterile:** Varieties 'A' and 'B' both fail to produce fruit with viable seed when pollinated by each other.

**Interstock:** It is a piece of stem inserted by means of two graft unions between the scion and the rootstock.

***In vitro* propagation:** Propagation of plants in controlled, artificial environment, using plastic or glass cultures vessels, aseptic techniques, and a defined growing medium.

***In vitro* transformation:** A heritable change, occurring in cells in culture, either intrinsically or from treatment with chemical carcinogens, oncogenic viruses, irradiation, etc., and leading to the acquisition of altered morphological, antigenic, neoplastic, proliferative, or other properties.

**Involucel:** The involucre of a secondary umbel.

**Involucre:** A series of bracts surrounding a flower cluster or some times a single flower.

**Irregular:** Used in botany to indicate a bilaterally symmetrical structure as for example, a bilaterally symmetrical flower.

**Isomerous:** With the members of the various series of flower parts of equal number.

**Keel:** A ridge along the outside of a fold, like the keel of a boat. The term is applied to the two coalescent lower petals of a papilionaceous corolla of the pea family.

**Labiate:** With lips, that is, two opposed structures as for example, the upper two petals as opposed to lower three petals in the mint family.

**Lanolin:** The natural fat of wool.

**Latent bud**: A bud, usually concealed, more than one year old, which may remain dormant indefinitely or may develop under certain conditions.

**Lateral bud:** One in a leaf axil.

**Legume:** A dry several seeded fruit formed from a single carpel and dehiscent on both margins.

**Locular:** Divided into locules.

**Loculicidal:** Dehiscent along the midrib of a carpel of an ovary containing more than one carpel and chamber.

**Lodicule:** One of the two small scales at the base of the grass flower.

**Loment:** A legume divided by constriction into a linear series of segments each containing one seed.

**Male gamete:** An antherozoid or a male gamete cell or nucleus.

**Male nucleous:** The male gamete of a flowering plant, that is, a single nucleus with the associated cytoplasm but with no cell wall.

**Megagametophyte:** A female gametophyte.

**Megasporangium:** A sporangium producing megaspore.

**Megaspore:** A larger type of spore developed by a plant with spores of two sizes or kinds.

**Megasporophyll:** A sporophyll bearing one or more megasporangia.

**Mericarp:** A portion of a fruit which appears to be a whole fruit.

**Mericloning:** A popular term, not in scientific usage, referring to the *in vitro* vegetative propagation of orchids from excised shoot tips, axillary buds, or floral organs.

**Meristem:** A plant tissue in which the cells are capable of, under favourable growth conditions usually in process of active division, to produce new daughter cells.

**Meristem culture:** *In vitro* culture of a generally shiny dome-like structure measuring less than 0.1 mm in length when excised, most often excised from the shoot apex.

**Meristemming:** A popular term, not in scientific usage, referring to the *in vitro* clonal propagation of plants from various explant sources including shoot tips, leaf section and calli; shoot apical meristems are seldom used.

**Meristemoid:** Meristem-like cells located in areas of a plant or culture other than the meristem.

**Mesocarp:** The middle layer of a pericarp.

**Metaxenia:** The supposed direct influence of the pollen or pollen parent on the characteristics of the developing fruit.

**Micropropagation:** This term is synonymous with *in vitro* propagation.

**Morphogenesis:** The process of growth and development of differentiated structures.

**Microspore:** The smaller type of spore in a plant producing spores of two sizes.

**Mixed bud:** A bud containing both leaf primordial and rudimentary flowers.

**Monoadelphous:** In one botherhood; referring to stamens with their filaments coalescent into a single tube.

**Monochlamydeous:** With a single series in the perianth that is only the sepals.

**Monoecious:** The stamens and pistils in separate flowers borne on the same individual.

**Naked bud:** A bud not covered by special scales but only by the outer leaves.

**Nocturnal:** Occuring at night. The term is applied to flowers which open at night.

**Nucellus:** The sporangium enclosed in the ovule of a seed plant.

**Nut:** A hard, relatively large, indehiscent, 1 seeded fruit.

**Nutlet:** Diminutive of nut.

**Nutrients:** Are defined as materials which supply either energy or essential mineral elements.

**Order:** A taxcon composed of related families or sometimes of a single family.

**Organ culture:** The maintenance or growth of organ primordial or the whole or parts of an organ in vitro in a way that may allow differentiation and preservation of the architecture and/or function.

**Organogenesis:** The evolution, from dissociated cells, of a structure which shows natural organ form or function or both.

**Osmosis:** The passage of water through a semi-permeable membrane.

**Ovary:** In the flower, that portion of the female organ which contains the potential seeds (ovules). After fertilization it grows into the fruit.

**Ovulate:** Producing ovules.

**Ovules:** The potential seed before fertilization.

**Panicle:** A cluster of associated spikes racemes or corymbs.

**Papilionaceous:** Butterfly like.

**Pappus:** The specialized calyx of members of the sunflower family; composed of bristles or scales.

**Parthenocarpy:** The development of edible fruit without fertilization. Parthenocarpic fruits are seedless.

**Parthenogenetic:** Developing without fertilization

**Partly inferior ovary:** With only the basal portion of the ovary adnate to the floral cup.

**Passage:** The transfer or transplantation of cells, with or without dilution, from one culture vessel to another. It is understood that any time cells are transferred from one vessel to another, a certain portion of he cells may be lost and, therefore, dilution of cells, whether deliberate or not, may occur. This term is synonymous with the term "subculture".

**Passage number:** The number of times the cells in the culture have been subcultured or passaged. In descriptions of this process, the ratio or dilution of the cells should be stated so that the relative cultural "age" can be ascertained.

**Peat:** It consists of remains of a aquatic, marsh, bog or swamp vegetation which has been preserved under water in a partially decompose state.

**Pectinate:** Resembling a comb.

**Pedicel:** The internode below a flower.

**Peduncle:** The stalk of a cluster of flowers or the next to the last internode below a single flower.

**Pentagynous:** With five pistils.

**Pentandrous:** With five stamens.

**Pepo:** A fruit of the type in the gourd family, that is, with a hard or leathery rind on the outside and fleshy placental tissue inside. The seeds are numerous and there is only one chamber.

**Perfect:** A flower having both functional pistils and functional stamens.

**Perennation:** A lasting state referring particularly to the persistence of fruit long after its usual season of maturity.

**Perianth:** A collective term for the calyx and the corolla.

**Perianth tube:** A tube formed by coalescence and adnation of the lower portions of the sepals and petals.

**Pericarp:** The wall of a matured ovary, that is, the wall of the fruit or the inner wall if the ovary is inferior.

**Perigynium:** A special sac which encloses the ovary of calyx.

**Perigynous:** Descriptive of the parts of a flower attached to a floral cup or tube which is not adnate to the ovary. A perigynous flower has a floral cup but the ovary is not adnate to it.

**Perigynous disc:** A disc adnate to the floral cup of a perigynous flower.

**Perlite:** It is a graywhite silicaceous material of volcanic origin, mined from larva flows.

**Permanent wilting point:** The amount of water in soils when rapidly growing plants fail to recover from wilting under condition of low transpiration.

**Petal:** One member of the series of flower parts forming the corolla.

**Petaloid:** Resembling a petal in colour and texture.

**Photosynthesis:** The process of combining in the presence of chlorophyll and light of carbon dioxide and water yielding glucose and oxygen.

**Photosynthetic:** Descriptive of a green portion of a plant capable of carrying on photosynthesis.

**Pistil:** The female organ of a flower.

**Pistillate:** Having pistils, that is, a flower which has pistils but no stamens.

**Placenta:** The tissue of an ovary which bears ovules or seeds.

**(Plant) Hormones:** (Synonym: phytohormones) are regulators produced by plants, which in low concentrations regulate plant physiological processes. Hormones usually move within the plants from a site of production to a site of action.

**(Plant) Regulators:** Are organic compounds, other than nutrients which in small amount promote, inhibit or otherwise modify any physiological process in plants.

**Plating efficiency:** The percentage of cells plated which give rise to colonies. The total number of cells in the inoculum, type of culture vessel and the environment conditions (medium, temperature, closed or open system, etc.) must always be stated. This term is often expressed as the percentage of individual cells in the vessel which give rise to colonies. If one is certain that each of the colonies arose from single cells, then one may properly apply the term "cloning efficiency".

**Pollen:** The spheroidal structures produced in an anther of a flower or in the microsporophyll of a gymnosperm. The pollen grains are microgametophytes developed from microspores.

**Pollen sac:** Pollen bearing cavity.

**Pollen tube:** A tube developed by a pollen grain and growing through the microphyle of an ovule where its contained cells or nuclei effect fertilization of the egg.

**Pollination:** The transfer of pollen to the stigma; or in a broad sense, the distribution of pollen. Pollination may be accomplished by insects, wind, gravity, water, birds and artificial methods derived by man.

**Pollinizer:** The variety (plant) used to furnish pollen. The male parent.

**Polyadelphous:** In several brotherhoods; with several groups of coalescent stamens.

**Polyandrous:** With an indefinite large number of stamens.

**Polyembryony:** Production of more than one embryo in a single ovule.

**Polygamo-dioecious:** With hermaphrodite and unisexual flowers on different individuals of the same species.

**Polygamo-monoecious:** Polygamous but in the main monoecious.

**Polygamous:** With bisexual and unisexual flowers on the same or different individuals of the species.

**Polygynous:** With a large indefinite number of pistils.

**Polymorphic:** With several or many forms.

**Polyploid:** With several to many times n chromosomes per cell.

**Population density:** The number of cells per unit area or volume of a culture vessel. Also, the number of cells per unit volume of medium in a suspension culture.

**Pome:** A fleshy fruit with several seed chambers, this formed from an inferior ovary, the fleshy tissue being largely the floral cup, the seeds not embedded in pulp. Apples and pears are the classical examples.

**Prickle:** A sharp, pointed outgrowth from the superficial tissues of the stem, that is, from the epidermis or the cortex, as in a rose prickle. The structure is not associated with the conducting tissues at the centre of the stem.

**Primary culture:** A culture started from cells, tissues or organs taken directly from organisms. A primary culture may be regarded as such until it is successfully subcultured for the first time. It then becomes "cell line".

**Protandry:** The pollen being discharged before the pistils are receptive.

**Protogyny:** The pistils being receptive before the anthers have ripe pollen.

**Pruinose:** Covered with a bloom, that is, finely divided particles of a waxy powder.

**Protoplast:** A plant cell without its outer retaining cell wall.

**Protoplast fusion:** Technique in which protoplasts are fused into a single cell.

**Pruning:** Means removing certain parts of the tree in order to modify and utilize its natural habits so that more and better fruits can be obtained at less cost over a longer period.

**Pumice:** Increases aeration and drainage in a propagation mix and can be used alone or mixed with peat moss.

**Raceme:** An inflorescence composed of pedicelled flowers arranged along an axis which elongates for an indefinite period. The lower flower blooms first and eventually the terminal bud forms the last flower.

**Racemiform:** In the form of a raceme.

**Rachis:** The axis of a pinnate leaf or of an inflorescence.

**Ray:** A pedicel within an umbel; a ray flower or its corolla.

**Ray flower:** In the sunflower family, one of the flowers with a ligulate corolla. The flattening is due to failure of complete coalescence between two petals of the sympetalous corolla.

**Receptacle:** The apical area beyond the pedicle, that is, the portion which bears flower parts. The receptacle consists of several or many nodes and short internodes.

**Regular:** Radially symmetrical.

**Reproductive bud:** A bud developing into a flower, cone or other pollen-, seed- or spore bearing structures.

**Resting bud:** An inactive bud.

**Rockwool:** This material is used as a rooting and growing medium.

**Rouging:** The removal of off-type plants or plants of other varieties.

**Rosette:** A condition in which the internodes are much shortened giving the leaves a bunched or clustered appearance.

**Salverform:** Descriptive of a sympetalous corolla with the slender basal tube abruptly expanded into a flat or saucer shaped upper portion.

**Samara:** A dry indehiscent fruit with a wing.

**Sand:** It consists of small rock particles 0.05-2.0 mm in diameter, formed as a result of weathering of various rocks, its mineral composition depending upon the type of rock.

**Scape:** A flowering stem which bears no leaves or only a small bract or a pair or whorl of bracts.

**Schizocarp:** A fruit which splits into one seeded sections (mericarp).

**Scion:** It is the short piece of detached, shoot containing several dormant buds which, when united with rootstock, comprises the upper portion of graft.

**Scorpioid:** Often used as descriptive of an inflorescence which is coiled in the bud stage as in the forget me not or fiddle neck. This is a helicoids cyme although it appears to be a spike or a raceme.

**Seed:** A mature ovule consisting of an integument (seed coat), an enclosed nucellus, the remains of the megagametophyte, the endosperm and embryo.

**Seed coat:** The outer coating of a seed developed from the integument of the ovule.

**Seed dormancy:** It is a condition where seeds will not germinate even when the environmental conditions (water, temperature and aeration) are permissive for germination. This not only prevents immediate germination but also regulates the time, conditions and place that germination will occur.

**Self-fertile:** The ability of a variety to produce fruit with viable seed following self-pollination.

**Self-pollination:** The transfer of pollen from the anthers of a flower to one variety to the stigma of a flower of the same variety.

**Self-sterile:** The inability of a variety to produce fruit with viable seed following self-pollination. (Some varieties may be self-fruitful, even though they are self-sterile, because of their ability to produce parthenocarpic fruits.).

**Self-unfruitful:** A variety which is unable to set and mature a commercial crop of fruit with its own pollen (or as a result of parthenocarpic fruit development).

**Sepal:** One of the flower parts of the outer series, the sepal forming a calyx.

**Septicidal:** Descriptive of a fruit dehiscent through the partitious between the seed chambers.

**Silique:** The elongate capsular fruit of the mustard family which has two seed chambers separated by a false partition from the middle of one placenta to the middle of other. The false partition in this case results from presence of some fertile and some sterile carpels in the pistil.

**Sod culture:** A method of orchard soil management in which a permanent perennial crop is grown between the trees. Mowed once or twice during growing season and then allowed to remain on the ground. A limited area around the tree is hoed, spaded or otherwise tilled.

**Sod mulch:** A method of charged soil management in which a permanent perennial crop is grown between the trees, mowed once or twice during the growing season and then allowed to remain on the ground.

**Somaclonal variation:** This type of genetic variation occurs when plant cells of certain species were grown in culture and new plant regenerated.

**Somatic cell hybrid:** The cell or plant resulting from the fusion of animal cells or plant protoplasts, respectively, derived from somatic cells which genetically.

**Somatic hybridization:** The *in vitro* fusion of animal cells or plant protoplasts derived from somatic cells which differ genetically.

**Spadix:** A spike with a fleshy or succulent axis, the flowers often partly embedded in the axis.

**Spathe:** A large bract enclosing an inflorescence at least when it is young. Spathes are either white or highly coloured but usually not green.

**Species:** A group of related varieties or often a single unit.

**Spike:** An inflorescence in which the sessile flowers are arranged along an axis. The basal flower blooms first, the last one formed is at the apex.

**Spur:** An elongated sac produced from a part of a flower as in the larkspur from a sepal or in the columbines from a petal.

**Stage I:** A step *in vitro* propagation characterized by the establishment of an asceptic tissue culture of a plant.

**Stage II:** A step *in vitro* propagation characterized by the rapid numerical increase of organs or other structures.

**Stage III:** A step *in vitro* propagation characterized by the preparation of propagule for successful transfer to soil, a process involving rooting of shoot cuttings, hardening of plants and initiating the change from the heterotrophic to the autotrophic state.

**Stage IV:** A step *in vitro* propagation characterized by the establishment in soil of a tissue culture derived plant, either after undergoing a Stage II pretransplant treatment or, in certain species, after the direct transfer of plants from Stage II into soil.

**Stamen:** The pollen-producing structure of a flowering plant, consisting of an anther and of a filament.

**Staminodium:** A sterile stamen, that is, without an anther or at least not producing pollen.

**Stem end:** The basal portion of an inferior ovary at fruiting time, the point of attachment of the pedicel.

**Stigma:** The apical portion of a pistil *i.e.*, the portion receptive to pollen. It is covered commonly with minute papillae and it is sticky through production of a sugar solution in which the pollen grains germinate.

**Stock or rootstock:** It is the lower portion of the graft which develops into the root system of the grafted plants.

**Style:** the tubular upper or middle part of a pistil connecting the stigma and the ovary.

**Subclass:** A taxon of a rank between class and order.

**Subculture:** See "passage". With plant cultures this is process by which the tissue or explant is first subdivided, then transferred into fresh culture medium.

**Subspecies:** A taxon of a rank between species and variety; a group of related varieties.

**Substrain:** A substrain can be derived from a strain by isolating a single cell or groups of cells having properties or markers not shared by all cells of the parent strain.

**Sucker:** A rhizome or a branch of a rhizome which comes to the surface and grows into a leafy shoot.

**Summer fallow:** Allowing land to lie idle for a season to conserve water for use the following season.

**Sunburn:** A killing of bark near the ground surface due to reflected heat rays, a form of sun scald.

**Suspension culture:** A type of culture in which cells, or aggregates of cells, multiply while suspended in liquid medium.

**Sterility:** The inability to set and mature fruit with viable seed. This failure may be due to non-function of the pollen, the ovule or both.

**Syconium:** An enlarged hollow receptacle bearing flowers and ultimately fruits inside.

**Sympetalous:** All the petals coalescent at least basally.

**Syncarp:** A structure composed of several fruits, these more or less coalescent as for example the fruits of pineapple or mulberry.

**Syncarpous:** With coalescent carpels.

**Syngenesious:** With anthers cohering in a circle.

**Synsepalous:** The sepals coalescent.

**Taxon:** A category used in classification as for example, a variety, a species, genus, family etc. Pural. taxa.

**Taxonomy:** The principles of classification.

**Tepal:** A term used for sepals and petals which are similar and not easily distinguished from each other.

**Terminal bud:** Bud at the end of the stem or a branch.

**Testa:** The seedcoat, that is, the hardened mature integument.

**Tetradynamous:** Having four long and two short stamens. A condition found almost throughout the mustard family.

**Tetragynous:** With four pistils.

**Throat:** The opening of a sympetalous corolla or a synsepalous calyx, that is, the expanding part between the proper tube and the limb.

**Thyrse:** A densely congested panicle, the main axis indeterminate, the lateral typically determinate and therefore cymose. A mixed inflorescence.

**Tissue:** Groups of cells having common functions.

**Tissue culture:** The term used for range of procedures used to maintain and grow plant tissues and organs in aseptic culture.

**Totipotency:** A cell characteristic in which the potential for forming all the cell types in the adult organism is retained.

**Transpiration:** The phenomenon of giving off water in form of vapours.

**Tuber:** A thickened short underground branch of the stem serving as a storage organ containing reserve food, ex potato.

**Umbel:** An inflorescence with the pedicles of the flower arising from approximately the same point. A compound umbel is umbal of umbels.

**Unfruitful:** A plant or variety that fails to set a commercial crop of fruit and mature it.

**Unisexual:** Of only one sex. Descriptive of a flower having only stamens or only pistils, not both, or of a gymnosperm producing only pollen or only ovules or of the gametophytes of pteridophyte bearing the reproductive organs of only one sex.

**Utricle:** A small, seeded more or less indehiscent fruit which appears to be inflated, that is, with a relatively thin pericarp more or less remote from the single seed. At maturity the utricle opens either irregularly or along a horizontal line.

**Vascular cambium:** It is a thin tissue located between the bark (periderm, cortex and phloem) and the wood (xylem).

**Vermiculite:** It is a micaceous mineral that expands markedly when heated. It is very light in weight, neutral in reaction with good buffering properties and insoluble in water.

**Vivipary:** The phenomenon in which seeds germinate in the fruit while still attached to the plant.

**Water berries:** A disorder of the grape in which the fruits are watery and fails to ripen properly.

**Wind burn:** A disorder of the leaves in which first their edges and later perhaps the entire leaf dries out and present a scorched appearance.

**Water spot:** A disorder of orange fruits characterized by water soaked appearance.

**Xenia:** The direct influence of foreign pollen on the part of the mother plant that develops into endosperm.

**Xyloporosis:** A disorder associated with lack of compatibility between stock and scion, characterized by pores or pits in the wood and corresponding pegs in the bark.

**Zygomorphic:** Bilaterally symmetrical. Often this is described as irregular, especially with reference to corollas.

**Zygote:** The fertilized egg, resulting from combining of the male and female gametes.

# References

American Association of Nurserymen. 1990. *American Standard for Nursery Stock,* AAN, Washington, D.C.

Beger, K.C. and Goug, E. 1939. Boron determination in soils and plants. *Ind. Eng. Chem.* Anal. Ed.II: 540-5.

Bhargava, B.S. 1999. Leaf analysis for diagnosing nutrients need in fruit crops. *Indian Horticulture,* 1 : 6-8.

Black, M. 1980/81. The role of endogenous hormones in germination and dormancy. *Israel J. Bot.,* 29: 181-92.

Bouyoucos, G.J. 1962. Hydrometer method improved for making particle size analysis of soils *Agron. J.* 54:464.

Chadha, K.L., Gill, H.S., Singh, B. Arora, J.S. and Wanjari, O.D. 1992. Report of mangament team on protected cultivation. I.C.A.R., New Delhi.

Chaudhary, S.M., Desai, U.T. and Kale, P.N. 1994. Effects of growth regulators on stooling in guava under semi arid conditions. *Journal of Maharashtra Agricultural Universities,* **19** (3) : 458-59.

Cheema, G.S., Bhat, S.S. and Naik, K.C. 1954. Commercial fruit of India. Mac Millan, London.

Czabator, F. 1962. Germination value: An index combining speed and completeness of pine seed germination. *For Sci.,* 8 : 386-96.

Dakshini, M.N. and Ratna Ramchandani; 2002. Taking Care of plants; Floriculture Today, **7** (5) : 7-13.

Dasberg, S. and Booster, E. 1985. Drip Irrigation Manual. International Irrigation Information Centre, Israel, 132 p.

Dikshini, M.N. and Ramchandani, R. 2002. Taking care of plants. *Floriculture Today*, 7(5): 7-13.

Flemion, F. 1938. A rapid method for determining the viability for dormant seeds. *Contrib. Boyce Thomp. Inst.* 9 : 339-51.

Grindal, E.W. 1990. Pot mixture, soil, fertilizer and manures. *Everyday Gardening In India*. Pp 147.

Hamrick, D. 1988. The covering choice. *Grower talks*, 51 (12): 64-74.

Harrington, J.F. 1963. The value of moisture resistant containers in vegetable seeds packaging. *Calif. Agr. Exp. Sta. Bul.*, 792: 1-23.

Hartman, H.T., Kester, D.E. and Davis, F.T., 1993. Anatomical and physiological basis of propagation by cutting in plant propagation: Principles and Practices: 209.

Hartmann, H.T., Kester, D.E. and Davies, F.T. Jr. 1993. Plant propagation principales and practices. 28 p.

Hildebrandth, C.A. 1987. Economical propagation structure for the small growers. *Proc. Inter. Plant Prop. Soc.*, 36 : 506-10.

Hodgson, R.W. 1961. Horticultural application of polyembryony in citrus. *Indian J. Hort.*, 18 (4) : 245-250.

Hudson, T., Hartmann, dale. E., Kester, Fred. T. and Davies; Jr. 1993. Propagation structures, media, fertilizers, sanitation and containers. Plant Propagation Principles and Practices. Pp. 28.

INCID, 1994. Drip Irrigation in India. Indian National Committee on Irrigation and Drainage, New Delhi, 176 pp.

International seed testing association (ISTA)., 1985. International rules for seed testing. *Seed Sci. and Tech.*, 13: 299-513.

Jackson, M.L. 1973. Soil chemical analyser. Prentice Hall of India Pvt. Ltd., New Delhi.

Kains, M.G. and Mcquestem, L.M. 1960. Propagation of plants. Orange Judd Publishing Company, Inc. New York, 2nd Ed. P. 222.

Kalra, Y.P. and Maynand, D.G. 1991. Methods manual for forest soil and plant analysis information report Nor-Y-319. Foresting Canada, North West Region northern Forestary centre, Canada.

Kogl, F., Haagen-Smit, A.J. and Erxleben 1934. Uber einneus Auxin (Hetereoauxin) aus Harn, *XI. Mitteilung. Z. Physoil. Chem.*, 228 : 90-103.

Kotowski, F. 1926. Temperature relations to germination of vegetable seeds. *Proc. Amer. Soc. Hort. Sci.*, 23 : 176-84.

Kyte, L. and Kleya, J. 1996. Plants for test tubes. An introduction to micropropagation. Timber press Oregon, 239p.

Lakon, G. 1949. The topographical tetrazolium method for determining the germinating capacity of seeds. *Plant Phys.*, 24 : 389-94.

Lindsay, W.L. and Norvell, W.A. 1978. Development of DTPA soil test for Zn, Fe, Mn and Cu, *Soil Sci. Soc. Am. J.*, 42: 421-48.

Mascarenchas, A.F. 1991. Handbook of Plant tissue culture. ICAR, New Delhi, 182 p.

Mishra, S.K. 1990. Effect of nitrogen and potassium on the performance of nursery plants of lemon (*Citrus limon*). Thesis Ph.D. (Horticulture) submitted to G.B. Pant Univ. of Agri. and Tech., Pantnagar

Motial, V.S. 1963. Polyembryonic studies in some lime, lemons and other citrus rootstock variety. *Sci. and Cult.*, 29 : 460-61.

Mukhopadhyay, T.P. and Sen, S.K. 1986. Guava. In : Propagation of Tropical and Subtropical Horticultural Crops. Bose, T.K. Mitra, S.K., Sadhu, M.K. and Das, P. (eds). *Naya Prakash, India*. Pp. 402.

Nijjar, G.S. 1972. Litchi cultivation. Punjab Agri. Univ., Ludhiana.

Page, A.L., Miller, R.H. and Keeney, D.R. 1982. Methods of soil analysis part 2 (Ed.) Non, Agronomy Services A S A – S S A. Publishers, Madison, Wisconsin, U.S.A.

Prasad, M.B. and Ravishanker. 1980. Studies of polyembryony and seed morphology in some important citrus rootstocks. *South Indian Hort.*, 30 (2) : 131-133.

Rathore, D.S. 1977. Stooling techniques in guava. *Prog. Hort.*, **9** (3) : 46.

Saroj, P.L. and Pathak, R.K. 1994. Propagation of wild guava through stooling. *Agril. Research in India.*, **2** : 74-80.

Shanker, G. and Rangnath, A.S. 1974. A quick test for Aonla, Ber and guava seeds viability. *Curr. Sci.*, 3 : 18-19. *Hort*

Sherry, W.J. 1986. Greenhouse covering materials: Optical, thermal and physical properties. *Grower talks*, 51 (9) : 48-57.

Singh, L. and Singh, R.N. 1955. Studies in the nuclear seedlings of some common rootstock of citrus. *Indian J. Hort.*, 12 (1): 53-59.

Thimann, K.V. 1935. On the plant growth hormone produced by *Rhizophus buinub. J. Bio. Chem.*, 109: 279-91.

Tiwari, R.B. and Tiwari, J.P. 1993. A note on the nutritional requirement of guava saplings. *Haryana J. Hort. Sci.*, 22 (3): 200-203.

Walkley, A.D. and Black, C.A. 1934. Estimation of soil organic carbon by the chromic-acid titration method. *Soil Sci.*, 37 : 29-38.

Ware, R. and Frankhauser, I.S. 1988. What should we cover the greenhouse with? *Proc. Inter. Plant Prop. Soc.*, 37 : 161-65.

Williams, C.H. and Steinbergs, A. 1959. Soil sulphur fractions as chemical indices of available sulphur in some Australian soils. *Australian. J. agric. Res.*, 10: 340-352.

Wood, J.S. 1985. Sunframe propagation. *Proc. Inter. Plant Prop. Soc.*, 34 : 306-11.

# Index


## A

Acid 28
Acricides 58
Aestivation 4
Age 4
Agrosan G.N., 65
Air-layering 34
Annuals 10, 104
Antibiotics 55, 58
Aromatic plants 13
Aromatics 3
Asexual 36
Avenue 11
Axe 1

## B

Balance 18
Balling 140
Bareroot 16
Basket 52
Beehive designs 108
Biennial 10
Blotting papers 21
Bonsai 76
Botanical name 4
Botanical names 6
Bouquet 96
Bouquets 104
Breaking of dormancy 124
Browning 137
Bubller irrigation 116
Budding and Grafting knives 1
Budding knife 36
Bulb crops 7
Bulbous Plants 74
Burlapping 140
Button Holes 96
Button holes 104

## C

Cactii 11
Can 1
Capillary irrigation 116

Carbon Source 133
Carpels 4
Cerasan 65
Chains 41
Check basin irrigation 116
Chip budding 39
Cleaning 135
Cleft 36
Climbers 11
Cold Frame 130
Cold Water Treatment 28
Cole crops 7
Common name 4
Completely Randomized design (CRD) 107
Compost 14
Condiments 3, 6, 8
Control of germination 124
Cotyledon 4
Covered petri dish test 20
Criss-cross (or Strip plot) designs 108
Crocks 68
Cucurbits 7
Cut Flower 104
Cutting 32
Cycades 11, 12


## D


Decoration 95, 96
Dibbing 84
Dibbler 1
Disc 4
Dissecting microscope 26
Draw 18
Drawing sheet 79
Drip irrigation 116


## E


Edge plants 10
Eraser 79
Excised embryo test 20
Explants 134


## F


F.Y.M 14
Family 6
Ferns 11
Fertigation 118
Flood irrigation 116
Floral display 104
Flower Bouquets 96
Flower show 102
Flowering 11
Flowering shrubs 10
Foliage Application 118
Foliage plant 104
Foliage shrubs 10
Forecep 26
Formal Gardens 79
Frost-Injury 86
Fruits 3
Fruits vegetables 7
Fungicides 58
Furrow irrigation 116


## G


Gajra 96
Garden 1
Garden tools 1
Gardens 79
Garland 96
Germination 20
Germination tests 20

Germination value 23
Grafted 4
Grafting 36
Grafting knife 32
Green house 98
Growth control 124
Growth Regulators 133
Gulkand 96
Gynoecium 4

**H**

Hand cultivator 1, 55
Hand dibbler 55
Hand Fork 1
Hand fork 55
Hand lens 3, 6
Hedge plants 10
Hedge shear 1
Herbaceous 34
Herbaceous border 87
Horticultural crops 3
Hot Beds 130
Hot Water Treatment 29

**I**

Identification 3
*In situ* 3
*In-vitro* cloning 132
Indocarmine test 20
Industry 96
Informal Gardens 80
Inhibition of abscission 124
Inorganic Salts 132
Insecticides 58
Irrigation 85

**J**

Judging 102

**K**

Khurpi 1, 14, 41, 58
Knife 3, 6

**L**

L disp 104
Label 63
Labels 21
Laboratory 3
Landscapes 3
Lathhouses 130
Lawn 82
Lawn grasses 12
Lawn mover 82
Layering 32
Layout 41
Leaf cutting 34
Leafy vegetables 7
Leguminous vegetables 7
Lifting 47
Liming 85
Locules 4
Lopper 1, 48

**M**

Maintenance 84
Mangroves 12
Marshyland plants 12
Mean daily germination (M D G) 23
Mean days 22
Measuring tape 14, 41
Medicinal plants 13
Medicinals 3
Metal tags 52
Microsprayer irrigation 116
Microsprinkler irrigation 116
Miscellaneous crops 8
Moss grass 32

Mother plant 32, 36
Mulching 65

N

Natural order 4
Neighbour – balanced designs 108
Nelder design 108
Notebook 63
Nurserybed 16

O

Orchids 12
Organic Supplements 133
Ornamental 3
Ornamental Plants 60
Ovary 4

P

Palms 11
Pandal 96
Paper bags 63
Patch budder 1
Patch budding 39
Peak value 23
Pegs 41
Pencil 6, 63, 79
Pinching 77
Plant Protector 1
Plantation crops 3
Planting board 41
Plastic strips 32, 36
Pole pruner 1
Poly house 98
Polyembryony 26
Pop-up irrigation 116
Position 4
Potting 68, 69
Potting Mixture 68
Potting mixture 92
Prevention of bud dormancy 124
Proliferation 136
Propagating Frames 130
Propagation 36, 124
Pruning 50, 72
Pruning knife 1, 48
Pruning saw 1, 48
Pruning shear 1, 48

R

Rake 1, 14
Randomized complete block design (RCBD) 108
Real value of seeds 24
Rectangular beds 16
Recurrent apomixis 26
Religious Use 95
Repotting 68, 70
Ring basin irrigation 116
Rock plants 13
Roecium 4
Rolled towel test 20
Rolling 85
Root Cutting 34
Root irrigator 1
Root vegetables 7
Rooted cuttings 17
Rootstock 4, 36
Rope 14, 58

S

Salad crops 8
Sandy loam 16
Scale 3, 18, 79
Scarification 29
Scion 36
Scissor 21

Scissors 63
Score Cards 103
Scraping 85
Secateur 1, 32, 36, 48
Seed germination 124
Seed Sowing 84
Seed viability 20
Seed weight 18
Seedbeds 14
Seedcoat 4
Seedling 4
Seedling growth 124
Seeds 18, 55
Semi hardwood 33, 34
Shield or T-budding 39
Shovel 1
Shrubs 10, 75
Side 36
Simple Ground Layering 35
Softwood 34
Spade 1, 14, 41
Spices 3, 6, 8
Spirit level 65
Spliced approached grafting 36
Split Plot design 108
Sprayer 58
Sprinkler irrigation 116
Sprit level 87, 90
Stair 48
Stigma 4
Stooling 138
Stratification 30
Style 4, 77
Subtropical 5
Succulents 11
Summer 10
Surface Application 118
Sweeping 85

## T

T T C test 20
Tally counter 21
Tape 3
Temperate 5
Testing of Plant Tissue 120
Thiram 65
Tongue 36
Top-dressing 85
Training 48, 72
Transplanting trowel 1
Trees 11, 75
Tropical 5
Trowel 68
Tuber crops 8
Turf grasses 12
Turf-plastering 84
Turfing 84

## V

Variety 4
Vegetable crops 6
Vegetables 3
Vegetative apomixis 26
Veneer grafting 36
Veni 96
Vernier callipers 3, 6
Vitamins 133

## W

Watch glass 26
Water plants 12
Watering can 58
Weedicides 55, 58

Weeding 85
Whip 36
Wild Garden 80
Winter 10
Wiring 77
Wooden poles 41

**Y**

Yield estimate 52